Story of a Khmer Rouge
Holocaust Survivor
and the
Creation of the
Kosol Ouch/David Lowrance
Rain Maker Device

Story of a Khmer Rouge Holocaust Survivor and the Creation of the Kosol Ouch/David Lowrance Rain Maker Device

By
Kosol Ouch, Rattana Keo Phuong, David Lowrance and David G. Dawson

E-BookTime LLC
Montgomery Alabama

Story of a Khmer Rouge Holocaust Survivor
and the
Creation of the Kosol Ouch/David Lowrance Rain Maker Device

Library of Congress Control Number: 2007940727

ISBN: 978-1-59824-731-2

First Edition
Published November 2007
E-BookTime, LLC
6598 Pumpkin Road
Montgomery, AL 36108
www.e-booktime.com

Thanks

To Nathan Bussa, Charle Oney (Chad), Micheal Hensdell and all of the top flight security staff members in the casino in Auburn. I am very pleased to acknowledge our friendship in this sixth book of mine. Thank you for your support and love in the creation of this book.

Khmer Rouge Survivor Story 9
The Khmer Rouge Story – Another Source 16
The Khmer Rouge Story – Historical Photos 20
Acknowledgments 34
Ionizational Machine 35
Experimental Record - Joe Cell Technology, and
    Torsion Fields 48
Motion 72
Joe Cell 87
A New Focus 100
Guardian Message 115
RainMaker Discussion 119
Group Discussion 131
Technical Overview RainMaker 1 Device 141
Magnetic Vortex Generator - Construction Detail 150
ZPE Coils 158
Helix Coils And Scalar Coils 183
Stargate Ascension Method.. 191
Kosol: Love Conquers All.. 193
Using Rainmaker Devices.. 194
Miscellaneous Pictures 198

Khmer Rouge Survivor Story

Hello all. Please join me, Kosol Ouch, and Rattana on this journey through the testimony of his experiences during the Khmer Rouge era. Rattana is one of the survivors of the Khmer Rouge holocaust project that was put on the Khmer people to experience in Cambodia from 1975 through 1979. Please join Kosol Ouch and Rattana as we both reveal to everyone the journey, drama, horrors and as well the survival of the courageous people that has survived the Khmer Rouge regime and era. So here is their story.

Kosol: Rattana, I will be your questioner and interviewer, can you tell me who are you, where you were born and your journey through the Khmer Rouge experience?

Rattana: My name is Rattana Keo Phuong. I was born on April 5, 1971 in Phnom Phen, the capital city of Cambodia, commute of Tholtompong Kan Pram Pert Mekara. First of all I was living during the Khmer Rouge time, my dad's name was Sarkorn Keo and he was one of Lon Nol army. He was attacked in near Phnom Phen international airport in April 10, 1975 by the Khmer Rouge army. One week before Phnom Phen was falling to Khmer Rouge army in April 17 1975 my dad asked my mom, Ane Phon Khvann, to go pick up money from the lieutenant commander of Lon Nol army headquarter and five minutes later pointed their AK 47 automatic rifle at my mom, brother, sister, and me and ordered us to leave the house and if we don't comply my family and I will be executed on the spot.

They told us to go to Cambodia country just for five day, and the Khmer Rouge told us to take a little bit of things with us, a few clothes and a little food, that they only needed for five day to

9

survive and then they will be authorized to come back to their home. The Khmer Rouge also told us and the people of Cambodia that the Americans are going to bomb Phnom Phen city tomorrow so they have to evacuate everyone from the city to get them to the countryside of Cambodia so they can survive the American bombing raid. I, Rattana, was four years of age during that time. So I had to live with the Khmer Rouge during the fall of Phnom Phen city. People had to escape during that time and exile to the countryside of Cambodia.

One year later I was 5 year of age in 1976. One of the Khmer Rouge leaders, Aung Ka (organizer leader), caused me to be put with one group of children of like age from 5 year of age to 13 years of age. There was 355 children in this group that I was in. So I was in this Khmer Rouge boy scout Kung Pol (group platoon) of number 71 regiment, city of Battambong commute of Phnom Srok Kun Ta Peng Thmor border of Siem Riep city Da Bong Pram. During that time all the children carried bamboo sticks for the weapons to protect the childrens army of Khmer Rouge. We worked 18 hour a day, over worked and exhausted by doing hard labor work such as digging ditches as well as picking up bones of dead corpses and burning them into ashes and turning the skeleton's ashes into fertilizer for the crops which was rice. Every night when we are sleeping, we have hut house which is open and is made from palm leaves that many Khmer Rouge boy scouts can sleep in. This hut house is not protected from wet weather conditions so many of the Khmer Rouge boy scouts got sick when they got wet, as well there is no medical treatment for any of the inhabitants of this open palm hut house.

The Khmer Rouge organizer Aunka separated us from our parents and brothers and sisters so we have to live with the Khmer Rouge army boy scout. The quality of life is very degrading, so they told us not to think about our biological parents, sister, brother, uncle, and grand parents, etc. The Khmer Rouge tell us to respect the Aunka leader only. The Khmer Rouge don't provide enough nutritional food for any individual at one time that is in the boy scout groups so this led to mass suicide. This led to cannibalism. The one who is living, they eat those who are dead and has committed suicide due to starvation. The dead bodies of those who killed themselves are fed to those who are living. Once a week Angka announced a meeting with all individuals in the group. All the top leaders of the Khmer Rouge army were Vietcong. The top leader of the Khmer Rouge are fluent in Cambodian and also they are fluent in the Vietnam language as well they changed their name from their native Vietnamese name to Cambodian name so the world wouldn't know the actual truth that the Vietnam communists are the orchestraters of the entire genocide of the Khmer civilization and people.

This is the truth behind the hidden truth of the communist Vietnamese regime. So the agenda of the Vietnamese regime was to wipe Cambodia off of the world's maps. The hierarchy of the Khmer Rouge army are Chinese. This Khmer Rouge hierarchy they speak fluently in Cambodian language. As well they also changed their name from their native Chinese name into Cambodia name. Their agenda is to wipe Cambodia and its people off from the face of the earth so they can split the land between the Vietnamese regime and Chinese regimes. As well to keep only 50 thousand Cambodian women only to survive. The rest of Cambodians both men and boys, will be executed. Both China and Vietnam grand plan is after exterminating all Cambodian people except the chosen 50 thousand women is to bring the men from China and Vietnam to reseed and colonize Cambodia's land with their off spring using the 50 thousand Cambodian women as surrogate parents.

The Chinese leader of the Khmer Rouge told the top commander of Vietcong that speak fluently in Cambodian to go ahead and execute the Cambodian people that is including men, women, boys, and girls except the 50 thousands chosen women. Then after the civilians are executed the next are the Khmer Rouge soldier and army including women, men, boy the whole nine yard are to be exterminated as well.

In 1977 the Angka top Khmer commander Vietcong told the children that they can not catch fish and grow vegetables on their property, that the Khmer Rouge will provide for all the basic needs. That means one spoon of rice a day, twice a day that is lunch and dinner. I have very difficult time to survive, I have to eat sweet potato leaves by boiling into hot pot of water. We boil banana tree by chopping it into little pieces and cooking so we can eat it as well extract all the nutrition that we can from it. At this point the quality of life is very very low, lower than animals. Once a month the Khmer Rouge lets us visit our biological parents and family. Then we have to go to the labor farms that are to dig dirt and ditches, from 4 am in the morning to 6 pm in the evening. We are really tired after the long days labor plus they only fed us two teaspoon of rice twice a day. So therefore the body is taxed from nutrition and energy, both physically, emotional, mentally as well as spiritually. These hard labors digging farms very harsh environment for anyone.

Kosol: So Rattana, these labor farms, do people die in these farms?

Rattana: Yes, there is too many people who died in this labor farms, the Khmer Rouge worked the worker force to death. If they don't work they don't eat. As well they can be executed right on the spot if they show signs of laziness or if they are sick and cannot work. Whoever was sick and cannot go to work, the Angka organizers will accuse you to be the enemy of the people and will execute right on the spot. The quality of life for these workers is severely unlivable.

Kosol: Who is Pol Pot? Is he the ring leader; or is the king of Cambodian the ring leader of the fall of Cambodian civilization? Can you explain?

Rattana: Pol Pot's real name is Sarl Lot Sor was born in Kompong Tom City. He was educated in Paris as red Khmer in France. Pol Pot was the protégé and learned from three different sources his paradigm way of thinking from Mao Tse-tung the Chinese leader of the revolutionary army.

Mao Tse-tung learned this paradigm way of thinking from Ho Chi Minh the Vietnamese army leader of the revolutionary army.

As well from the Paris Revolutionary Army in France. Then he began to put into this practical application of what he learned from this three source of paradigm by putting into practice in Cambodian country and its people.

Pol Pot is the 2nd leader of the Khmer Rouge regime and the first leader is the king of Cambodian his name is King Sihanouk, the third leader of the Khmer Rouge regime is Kel Som Purn, the fourth leader of the Khmer Rouge regime is Eng Sary, the fifth leader of the Khmer Rouge regime is Tah Mok and sixth and final leader of the Khmer Rouge is Noun Chea.

King Sihanouk was a playboy when was a prince, he slept with lots of women who take care of the Phnom Phen palace. He was a song lover and he liked to have sexual and pleasurable fun with young women at that time. As well he doesn't take care of the country and the people's well being. Sihanouk was more focused on being a playboy or Casanova than being a king. Sihanouk's ideology was communism so he enforced his loyal subject that is in his courts to follow and orchestrated that particular communism doctrine. Sihanouk told Pol Pot (Sal Lot Sar) Kel Som Purn, Eng Sary, Tah Mok, and Noun Chea to take his communism idea to combat Lon Nol republican army to negate America from supporting Cambodian civilizations and Lon Nol and his republican army.

From 1970 through 1975 Lon Nol was supported and furnished weapons from the United States of America. Once Sihanouk completed his take over of Cambodia from Lon Nol, then Pol Pot put his own agenda into play by putting his regime leader King Sihanouk in house arrest. Then Pol Pot became the ultimate leader of the Khmer Rouge regime from 1975 through 1979. After the Vietnamese army invasion of Cambodia, then they battled the Khmer Rouge army and pushed them to the border of Thailand next to the refugee camps.

That is where they stayed to this very day. Now they are just being integrated into the current developement civilizations of the Cambodians. After the Vietnamese invasion of 1980 the Vietnamese army put 10 million land mines in Cambodia to kill all the Cambodian people. That is one land mine for each person, Cambodia that time has only 10 million people, when Vietnam army came to Cambodia and took everything from the Khmer Rouge people and Cambodians including weaponry, vehicles, etc., back with them to Vietnam. In other words, pure robbery and raping of land, culture, technology as well as people.

In that time Vietnam put Heng Somren to be the president of Cambodian as a puppet so they can tell him what to do, and when to do it from 1979 to 1985. After that the Vietnamese made a declaration for Heng Somren to be removed from power. Then King Hun San was appointed by the Vietnamese leader (who oversaw the caretaker-ship of the Cambodian land and her people) to be Prime Minister from 1985 until present time 2007 and beyond. Cambodia right now is just starting from the down fall of the Khmer civilization and rising to the present technological and educational and as well the rising economy to be in the world's markets of the present time.

Kosol: What have Cambodia and you learned from the rise and fall of both the Khmer civilization and the rise and fall of the Khmer Rouge era?

Rattana: So transformed from a very simple minded people to a very conscious society so I want this to be a written record for the world to see and read that the Cambodian people are a resilient people and I myself am proud to be one of them.

I know there are many challenges for the survivors of the Khmer Rouge era and their children all over the world because during that 4 years of Khmer Rouge regime rule 1.7 million Cambodian people were sentenced to death by execution. They were sent to death camps like, for example, S-21, which was a high school (sa la toh slang prisoners) in Phnom Phen that was converted into a death and torture camp where many both men, women, and children was sent to be tortured and also to be executed as part of the Chinese and Vietnamese orchestrated agenda to further advance the agenda of the extermination of the Khmer people during the Khmer Rouge regime that is orchestrated by the Vietcong and the Republic of China who sent their spies and infiltrators to help and orchestrate the expansion of the

Chinese and Vietnamese interests by exterminating the Khmer people and her civilization.

To this end, now the hidden agenda of China and Vietnam toward Cambodia and her people has come to light. The world can read and make up their own minds of how and why the history of the Cambodian living the Khmer Rouge era went down. May the future generations of both the survivors of the Khmer Rouge era and the world never forget what happened to their parents and grand parents that was experiencing the Khmer Rouge era may they never forget where they come from and are going may this history aspect of the Khmer people live in the heart of all Khmer people all over the world so the next generations never forget what the Khmer people have been through, so that the dark history of the Cambodia people won't repeat itself in future generations.

Kosol: Rattana and I wanted this current generation that is living right now to read and understand that what was going on in the past is part of their history and the history of their civilization. Both the dark and the goods is all part of their heritage. Both Rattana and I don't want any future generations to experience what the past generations have experienced.

The Khmer Rouge Story – Another Source

To continue on with the Khmer story from another source who is very knowledgeable about the era as well. He will be called Source.

Kosol: Hello Source. Can you elaborate on what was going on with Cambodian events in the Khmer Rouge era?

Source: When the communist Cambodians took over the country in 1975 most Cambodian people were celebrating, because we are hoping finally the peace to come to Cambodians. Some of the Cambodians shaved their hair, and some were kissing the ground and most thanked the lord (God) for bringing peace to Cambodians. Three days after the occupation most of the Cambodian people have a night mare because the communists forced all the population from the city to move out to avoid the American bombing of the city.

For the whole movement only one traffic getting out of the city. With all kinds of panic because we were forced by gun point by the new regime, they told us not to bring anything with us because we will back home in one week. For most of the Cambodian people who have transportation like truck, car, motorcycle, etc., were confiscated by the new Angka (organizer leaders) so we all ended up with nothing just walking for day and night with bare feet. Some were forced to keep moving to different local and countryside villages.

Most of us have to find residence to start new life, first all of us got to register with the new regime. Then they divided us into many different groups. After that all of us got to turn in our personal property, any things of value like cloth, jewelry, pots, basic items for survival were turned in to the Angka. Then they grouped us into different groups, for example, men 16 to 35 living in one group, and women also. The children from 9 to 15 living in one group away from their birth and biological family.

After that we all were assigned different projects by the Angka. Everyone was forced into hard labor work force. From early morning to evening, with little rest and little basic sustaining food and water. One month after this new regime took over this new country, they completely have control over the whole population of Cambodians. No one can move from village to village or visit any relative. And they started to screen the people who used to work for the former government as top level people such as police commissioner, governor, supreme judge, in other words high ranking official officers. They had lied to the people that they registered with the new regime so they can get theirs job back and met with the King (Somdach Sihanouk). Most of the people were so excited to join the new government and see the king but their excitement was shorted live, because most of them were the first groups to be executed by the new regime. Now everyone in the village started a new life with the new forced labor of hard work. The communist regime told all of us that we all will have a bright future so we all got to work hard.

No one will have any question about the government's policies. Everyone will keep working hard; even the young children also will participate in these policies from the government. Most of the Cambodian people kept working hard for the new government and the government keeps promising all kinds of positive future but in reality nothings really came true that they promised the people only death and hard labor forced work.

Day and night the people worked hard labor for this new government, because everyone was afraid to die. Everyone was told to keep any eye on each other, and no one is to trust each other, because it was only the Angka that people needed to trust. Angka told everyone that they are their parents, mom, dad, grandpa, etc., to all the people as well they psychologically told everyone they have a million eyes that they see everyone at all times so that made no one trust no one even with their own biological relative like mom, dad, children, aunt, and uncle as well brother and sister, don't even trust each other during this dark period under the communist government.

The government started searching and screening the people for those that didn't belong in their regime such as teachers, soldiers, doctors, government officials of the former regime. As they screen and search for these people, first they used rumor tactic started from their spies and used on to the people by starting rumors. Once this rumor reached its targeted person they immediately pulled the targeted person or group of persons away from their family to be executed right on the spot.

Even if you know you are going to die no one will attempt to run a way or escape from their foreseen death because if they tried to run away, then the Angka would then proceed to execute that person's entire family and related family. So therefore they was to die or else if they tried to run then their entire family will take their place.

When the government came to get you, you have no right to ask question why? The end result is you will die, by accusation from some unknown individuals or you will be executed for your biological geno makeup such as lighter skin, then they are accusing you are a rich people so therefore you will be executed. As you can see this regime has no positive view and outlook for their people who they governing because the people from the new regime were uneducated people so they were threatened by educated people, so therefore all educated people were their enemy.

Source: Now at the last day of the communist era in Cambodia when the Vietnamese army came into Cambodia, the Vietnamese army set up surgical strategic traps for Cambodian to kill Cambodian. They lured the civilian population of Cambodian to come into their controlled area, then they completely pulled out from their controlled area and left all the innocent civilian population that is in their controlled areas to be murdered by the communist Khmer Rouge soldiers, because the Khmer Rouge thought all the innocent civilians were sympathetic to the

Vietnamese army. Therefore they are considered to be traitors, that meant they would be killed on sight. So this military tactic was used to murder Cambodian population all over the country in 1979.

Kosol: Thanks Source for your insightful testimony that will show more light in to what happened in this dark era of the Cambodians experience of that era.

The Khmer Rouge Story – Historical Photos

Acknowledgments

I and the co-authors acknowledge Atmospheric Ionization Research and also Peter Steven the chairman of Atmospheric Ionization Research, Incorporated association 266 Molesworth Street Lismore 2480 New South Wales Australia who has given the author Kosol Ouch and co-authors pictures to be used in the book. I thank Peter Stevens and his atmospheric ionizations research organization for granting us the full permission to put their device in this book.

Ionizational Machine

Now as for the ionizational machine it consists of two Joe Cell system, a octahedron formation mirror and 8 copper siniod 12 gauge coils aligned in a octagon formation also has a Jack Toyer Base machine with the Joe Cell system, a garnet crystal (dedecahedron crystal lattice molecule structure) of many variety and garnet ruby (dedecahedron formations molecule structure lattice), mind links consciousness from the human operators intents consciousness links. Basically, that is it.

RainMaker device can be substituted for the Joe Cell device, or vice versa.

My friend Peter Steven uses similar consciousness principal to create the ionizations just as my RainMaker device so Steven has concluded that the Kosol RainMaker device and the ionizations is one and the same device because both device use mind consciousness intent links, torsional generators, crystal garnets and mine use quartz because I can't afford a garnet crystal ruby, all work in the same consciousness principal. No limit whatsoever.

Take a look at picture.

As well the Joe Cell and the RainMaker is called a torsional generator. The ionization RainMaker device integrated the Kosol RainMaker device or the Joe Cell device as it is the heart of the system and also the garnet crystal, garnet crystal has dedecahedron platonic crystal lattice structure that is very good for mind links when put over the Kosol RainMaker device or the Joe Cell device.

Now for running car with the Kosol RainMaker device, all you have to do is put the Kosol RainMaker into the aluminum metal container, or stainless steel container or copper container, etc.

35

Then drill a hole on top of the RainMaker metal container, connect this hole with a copper tube or aluminum tube either way but copper tube is preferred, then connect this tube to the carburetor of the car engine. Now remember you can also put a garnet crystal in the RainMaker device instead of the quartz crystals if you can afford garnet crystals that is. I believe it is a little more expensive then a quartz crystal you can get it online or at the super mall in your surrounding area at any mineral store dealer in the malls.

Now the torsional energy fields will travel from the RainMaker device to the metal container that contains the RainMaker device then from the container to the copper tube and the torsional energy will travel from the copper tube to the car carburetor from there it travels to the energy to allow the car to run from torsional energy only so your car don't have to run on gas anymore. No limit whatsoever.

The most important thing is that the garnet crystal is the dedecahedron crystal lattice that is the right structure lattice for the Kosol RainMaker device and the Joe Cell device as well the Peter Stevens ionizational RainMaker machine all of these are consciousness torsional generators.

Regards,
Kosol Ouch

Now from David G. Dawson:

RainMaker1 - Oz Weather

Have taken Amethyst Sphere off top of RainMaker and loaded one of the triple Bismuth Scalars into the top of the top Ferrite and packed around this with 8 x 1" Garnets.

Still have light rain here but see more on the way.

This event is unprecedented - the rain is still coming in directly up my tail - Deep Troughs have clearly formed with extensive Thunder Storms much earlier than usual with Spring oncoming.

Northern Territory, Queensland and SA have all had some good rains.

The Trough Line has been anchored from me here since beginning work with the PIG (Positive Ion Generator) on the 13th.

David L,

Yes, we have observed stagmentation of our systems by leaving in the one mode and a regular pulsing in the slow charge/rapid discharge mode is the answer or another device that may harmonize with the RainMaker and create a new condition of instability.

None of this was forecast!

There is a Low circulating over Mount Gambier in SA and the Low to my SE is still rotating towards NZ.

PIG is still pointing SE.

Kosol,

PIG - this is a 90mm x 2m Tube device made about 3 years ago but never trialed. Output is 5kV and works from 240v Mains - 'Heat' can be felt out of the end of the Tube but there is no actual heat from the electronics.

This technology is based on the principles of Lee Crock at 'The School Of Universal Energy'.

The disappearance of the Radar at times is due to its other use as a wind recorder.

'Weatherzone' has killed some of my 'tools' for Weather monitoring but hope these work for you.

David,

Yes, they look cubic but it seems to say in general or 'often displays' this form but not always when referring to Garnets.

There is a huge amount of information on Crystal structure and shape.

Garnets are the recognized Crystal for the Dodecahedron shape.

From my Crystal Lady's data - Stone Of Health - extracting negative energy from the chakras and transmuting the energy to beneficial use. Stimulates both base and crown chakras to allow kundahlini flow. Protective and calming. Commitment to a purpose.

From another source - Constancy, felicity, true friendship.

Yes, you are correct - an ability to connect to communicate seems the purpose here.
As I said, my 1" Garnets are tumbled but still hold a 5 sided appearance, no matter at what side you look from.

David,
PIG is simply a + Ion Generator which is isolated and suspended within the base of the 90mm tube.
No metals BUT the Silicon Diodes and the High V Capacitors.
This is a bit like an 'Orgone' projector of Wilhelm Reich.
There is a continual flow of air in from the base through the tube.
I have a feeling that there is an Aetheric attraction manifesting here and the tube projects the +ions plus the, probably, Tempic field (Orgone).
I must assume that 'spin' is being involved here also.
A Negative Ion Generator (I have its opposite right alongside me here, 2' long in same 90mm pipe, 1 metre away and pumping all the time, 24 hours) is said to promote well being and also reduce smells in the air - basically very healthy.

Me talking here - What I believe is happening is that Negative Ions allow for the concentration of Aetheric Energy in the vicinity of the emitting device (5kV).
This would make Negative Ions = Female or Implosive or Inflow and is CW in rotation.
In the PIG case we are throwing the + Ions out in an explosive or projective manner.

Positive Ions = Male or Explosive or Outflow and is CCW in rotation.

So basically we are projecting our + Ions plus the Tempic field out into the Atmosphere where it is probably attracting 'Life' in rainfall as an area of concentrated Aetheric Energy.

I basically see Water as 'Life' as a concentration of Aetheric Energy.

I have yet to see anybody explain Ions in this manner and it's good for me to be able to answer your question and at the same time put down on paper what I expect is really happening here.

I see the same with my TJ Constable Rotating Cone device (CRC) in the way it attracts a Trough to form which will fill with rain for disposal across the land. Slowing down the natural Aetheric flows so that they fill with 'Life' in the form of the Trough.

My pointing here is NW to SE and rotate CCW looking to the SW. CW has always given me fine Weather.

Our natural Aetheric flow over Oz is from SW to NE.

I will post the circuit for the Ion Generators and put all 3 together for a photo shoot and post.

Will also do the Garnets.

Appreciate answering the questions as it allows me to record what I have always felt were the true principles involved with these devices.

Thanks.

Smokey

Now the latest thing is the flower RainMaker. It has a container that looks like and opens and closes like a flower, and all you have to do is put the RainMaker torsional generators inside, then you can close and open it at will since the torsional field generated by the RainMaker can be contained. The container can be created with copper panel on the inside and sandwiching a dielectric paint or tape then on the outside panel is the aluminum panel. So both the copper panel and aluminum panel sandwich the dielectric material such as electrical tape or dielectric paint etc.

Yep, let the RainMaker generate the torsional field inside this flower like container for two to three hours then after that you open up the petal of the container a powerful blast of torsional field will emit and rush out of this container that is generated by the RainMaker device being contained by the capacitors flower like aluminum/dielectric/copper containers.

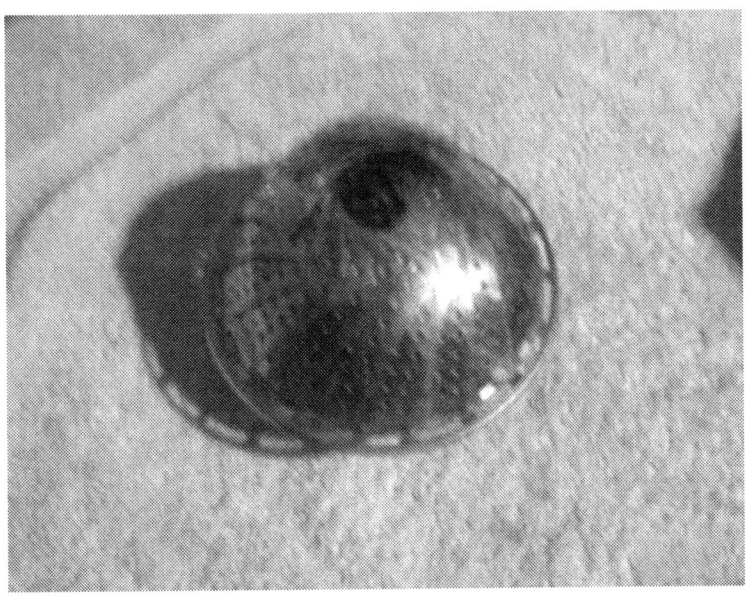

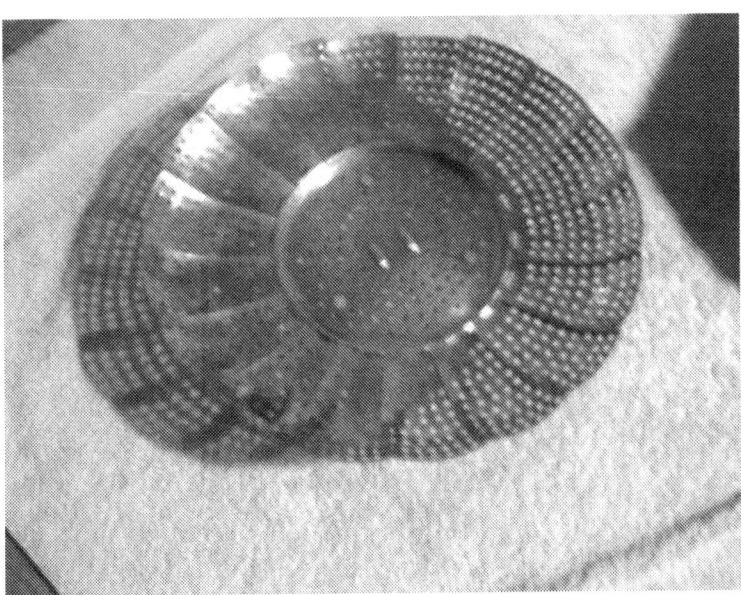

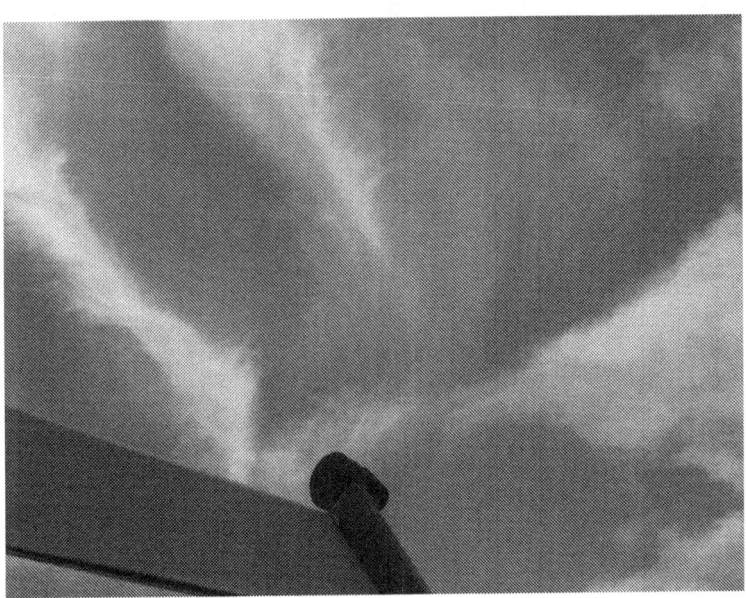

Experimental Record - Joe Cell Technology, and Torsion Fields

Bernie has supplied two Joe Cells for experiment of the c_s_s_p group. Dell and Dave have been experimenting with these and studying the torsional aspects of the cell at present. It is not yet known what the torsional function of the water does to an engine, or if the gas produced from the cell is also a factor. The experiment thus far deals only with the torsional component, which we are presently very familiar with.

Discoveries on the Joe Cell:

Voltage alters surface tension and viscosity of the water, this alters the spin frequency of the water and direction, and can allow for tuning the cell to various resonant torsional peaks as though it were a scalar torsion coil. Between .3 and 40 volts a range of resonant torsion peaks are discovered to exist. 1.5, 10, 24, 33 volts for this cell. Joe powered the cells negative lead to the inside of the cells and positive to the outer case. This causes the water to spin up the same direction as its natural earth spin state only at higher velocity, but only works this way in the southern hemisphere. The water is found to start giving up its iron deposits. Further the stainless steel will begin to leach chromium into the water which will grab up oxygen atoms to form $Cr\,O3$. The outside layer of stainless steel is in fact a layer of this material and why it resists corrosion in an oxygen rich atmosphere. The water will take on a green tint as the chromium slowly saturates it with Oxygen rich $Cr\,O3$. The hydrogen will bubble up and the Oxygen will become reactive inside the water, less escaping, during phase conditioning stages of the cell.

As the water is conditioned and filtered to smaller and smaller natural looping chains, clumps are removed, the torsion state becomes higher. The water becomes finer and drops all its soluble iron, possibly other metals as well. The water will ultimately run through a filter very quickly as it becomes more conditioned.

Magnetic Ring Experiment

To investigate the torsional conditions of the energy in the Joe Cell this experiment was performed as a comparison to the RainMaker vortex generator.

Joe Cell mod tried today 7 - 14 - 2007:

I built up a ring with iron tie wire and black tape, to place a ring of magnets at a 45 degree angle around the water line of the Joe Cell. This is direct application of the RainMaker discoveries and does in fact have a strong effect on the unit. [See pictures below.]

From RM experiment and tube devices we have learned:

Iron and 45 degree magnet rings will excite a diamagnetic field to expand outwards and become interactively connected. "North out" strong enough, will drive the system into outflow.

Torsion propagates through a magnetic field, between the AG metals which it freely flows down. Adding a magnetic field between two objects will cause the torsion to jump a gap. [See reference note.]

[Torsion can propagate down a magnetic wire surrounded with copper. It will also propagate down a copper tube surrounded with aluminum. These are ideal methods to couple torsion to other places away from the field generators.] This suggests a new method of coupling to an iron engine block as well.

Using this approach I made a discovery of note. I was attempting to push the Joe Cell into an outflow mode to stop the Rain here, as it is not a popular thing to do. The unit is observed to create inflow with either voltage polarity. I placed the familiar North out magnets aligning the top edges with the water line. A ring with 8 stacks of neos. Torsion from the magnets now intersects the water line at 45 degrees approx, and we see the exact same interaction as with the bismuth. The frequency of the torsion field increases, and now I can produce a large torsion field with only .3 volts of E field on the cell. The sky is already clearing up. I definitely have an outflow in effect at the magnets, but along the entire surface of the cell I am not certain. Testing with finger tips reveals a strong outflow off the magnets ends. This was a necessary step for me here in the rain belt of Alaska.

Off the top the field is almost too intense to hold the hands over at 33 volts. The resonant voltage points have moved and there is a very strong field at 8.8 volts. The field is more interactive as well with the mind in this mode. This is the basis of a new torsion field design that may use less energy and yet offer a simpler frequency control. No function generator is needed.

With the tube devices I was never able to get into outflow mode and rain was all I could get from them. By combining the JC with the RM I have gotten a very strong outflow device.

Note:

After this it was discovered to reverse the cells electric polarity is enough to drive the cell into outflow mode and reverse the spin field, in the Northern Hemisphere, and creates far less local turbulence to the environment or the weather. The magnet ring was dropped as not effective for coupling the energy into an engine. It did demonstrate that the energy working in the cell is the same energy as in the RainMaker and density spheres. It is a form of conscious energy.

Spin Fields in Water:

Ordinary water is placed in a plastic or clear container with a sealed lid. Sharply shake the water and note the bubble spin direction after stopping the container. It will spin slightly CCW [counter clock wise], in Alaska [Northern Hemisphere]. This is also observed in watching a stream of water being poured, the water will braid itself as it falls and spin CCW. Slowly pouring the water from a plastic container into a stainless steel bowl has also been used to detect spin direction successfully and this is an art form at present.

Picture showing the CCW spin on water from an artesian well in Sitka Alaska.

Now place charged water from the Joe Cell in the bottle and shake, or use any other method you have had success with. The bubbles will spin the opposite direction and at a much higher velocity. Spin direction is CW [clock wise] looking down from the top. The Joe Cell, as Joe built it, has reversed the waters background spin field at the nucleus of the atoms where its major mass sets. While the water normally appears motionless, this spin momentum is setting inside the water as an alternate charged potential. A [torsion force.]

Spin Field Direction:

While the Joe Cell was designed for Southern hemisphere spin direction, our experiments have led us to the following conclusions.

In the Northern Hemisphere water spins CCW naturally. To accomplish this with the Joe Cells reverse the voltage during the conditioning phase. Dell was the first to observe this at the c_s_s_p site and then it was further confirmed by myself. Positive lead to the center tube and negative on the outer tube. This will cause the water to increase its CCW spin matching the earth and the earths spin field will then keep the cell charged. While using water with a CW spin in the Northern hemisphere I have observed turbulence with the earth's fields. While this could be used to release a stuck weather pattern, its use would be temporary and then reversed to a CCW spin to match the earths natural spin field afterwards. The alternative is to draw in weather as shown below in the photos of the fog banks that rolled in.

45 Degree Magnet Ring At Waters Surface:

Charging water is not about producing bubbles, as Bernie has indicated. The presence of the magnetic ring lowers the waters ability to break down and increases the voltage that it takes to make bubbles. However it also alters the waters spin at a higher level, best guess, because the torsion fields are also increased. The magnet ring diverts the torsion out the ends of the magnets and radiates it to the sides, where hands can more easily feel the energy. The resulting field residue after the ring is removed is not worth it for this test however if one has been able to sense the waters natural spin torsion field using hands around the container. Degaussing the steel container may be necessary after removal of this experimental device.

I have reversed to South pole out to see if we get a similar effect that we do from the RM units. I also filtered the second batch of water and now have almost one gallon available for experiment. [6 hours later the area became covered with a heavy fog, and the weather was sucked in from the ocean indicating a strong inflow was present once again.]

Identification Of The Spin Charge Storage Material:

Removing the water from the cell it becomes apparent that the cell is not holding the charge, but the water is. The water is producing the same properties as a scalar canceling coil.

Integration:

These observations have direct link to the two devices and demonstrate that the cell can be used with very small voltages if magnets are added in the correct alignments.

The Joe Cell design gives us one more method of controls, that of a voltage reference to set vibrational frequency of the water.

I do not know if this will improve Joe Cells, however it does show that we are working with the same energy, and can incorporate the two into a common device. Since water is diamagnetic, and the E field can control the balance this had to be the case, but for me now, this places water into a new category of vortex generators.

Bernie, I believe your insights on getting us all together has paid off. Thank you. We now have a new core material to experiment with for our torsion systems.

Reference Note:

Torsion will jump from iron to copper to iron if a magnetic field is present. The function of the engine spark is probably providing this coupling as a pulsed connection. The torsion is jumping from the Aluminum tube to the engine during the arc when a magnetic field is present. A copper or Aluminum tube could be run closer to the spark gap to see if a stronger coupling is achieved.

Observed Weather Considerations

My Joe Cell experiments were conducted inside a very large steel building, and the effects have thus been very radiant. The above picture on the left was after one night of magnets arranged with South poles outwards, and conventional electric polaritys Joe used [negative lead to center tube]. The one on the right about 2 hours after reversing the magnets. The weather effects from the Cells geometry seem to be low altitude effects and have now appeared to produced very heavy fog layers on two occasions, [abnormally thick fog]. The fog blanketed the whole town to about 2 miles out where the sky was clear. Clearing the fog was very dramatic and began around the building first then spreading outwards.

Sitka is a very good place to observe these kinds of effects because of the high moisture content here, and the weather can often change almost instantly.

Magnet Ring

Picture showing the iron magnet ring devised to simulate the RainMaker structure around the diamagnetic water of the Joe Cell. The cell can be degaussed after this experiment for reuse as a Joe Cell. An AC field is applied then slowly turned down or backed off so as not to create any square waves during is shut down.

Charging Water

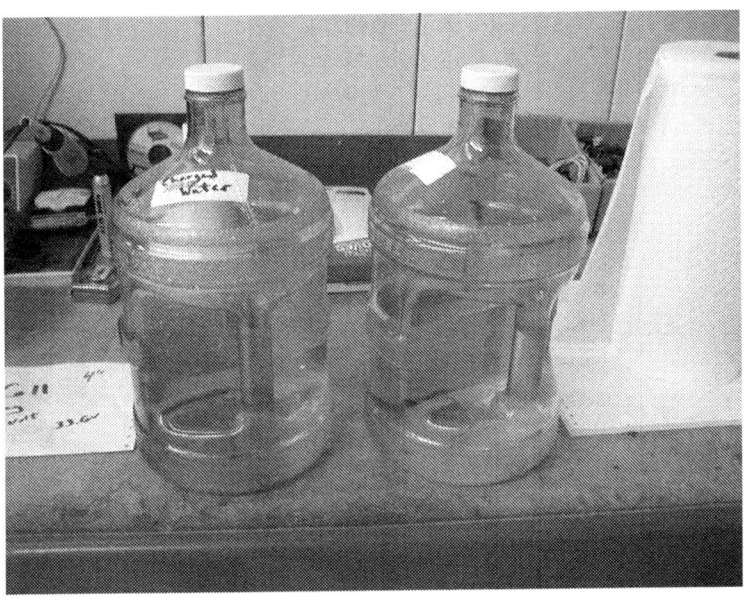

Picture showing the water after charging. Water on the right is normal spring water, on the left after charging radiates a torsion field. The green tint is normal and reflects the chemistry happening in the cell [chromium oxide release from the SS outer layer]. If you shake these bottles strongly you will see the bubbles shoot opposite directions and actually spin around the whole container in the charged water.

Torsion Forces

If you look closely at a cylinder on an engine you will notice similarities to the Joe Cell itself. Two iron layers with a water jacket between them and a spark of High Voltage at the center of the system. It was reported that head gaskets that are conductive stop the cell from working on the engine so it is likely that EM is a major factor, as well when the plug wires are all removed the engines stop. Replacing the gaskets with a dual layer of insulated gasket material enable the engines to work from the cell. Electric arcs produce a spiral spin when they jump. This sends out a torsion force as well as discharging the EM voltage gradient extremely fast. The spirals direction is a function of which direction the arc is moving. Electrons jumping towards the head spin opposite as jumping towards the piston. Also during the release of a HV gradient all the spin motions of the atoms setting inside the field drop their torsion gradient rapidly. This can be shown using two pennies and a small slug of Bismuth inside insulated on one side using electrical tape. Place about 60 V dc across the pennies and you will feel a constant torsion field come up out of the bismuth. Any diamagnetic materials setting inside this field will do the same, water as well, and you can feel this torsionally. Bismuth being the easiest to perceive and at 60 v it was very strong. The Joe Cell using water and stainless steel to do this also.

The reason a voltage gradient causes an increase in density is that it pulls the Proton layer and the Electron layer opposite directions. This causes them to move off center on there precession spin axles and narrow the gap on one side. As the opposing spin field [tempic vectors] move closer together they accelerate the time flow rate, and all atoms spin fields increase their c velocity. With the Bismuth experiment the field is not directional and seems to radiate all directions from the bismuth creating a density sphere effect. Using a magnet off one side you can direct the field by aligning more spin one direction, and the field will concentrate 90 degrees to the E vector and the B field vector.

Dowsing The Joe Cell System:

Angles off the top of the cell were detected using a copper tube 2 foot with aluminum tube over the lower section. The tube is held on the copper area which is radiant and torsion fields will light up the whole body. Between the insulated layers there is a strong heat energy felt with a finger as the two diamagnetic materials interact differently to the field. When held at the correct angles, and at the correct distances from the cells the tubes light up with Chi type heat energy, and can also be felt at the third eye center, exactly like with the RainMaker units. The torsion fields expand to fill the entire tube length and detection is much easier. I detected 3 angles at the top of the cells, around 66 degrees and 45 degrees and one somewhat lower. This would be where the upper cone should be designed I would guess. Also vertical off the center of the tube the copper will pick up all the 45 degree torsion angles from all the water layers in the cell and very hot fields were found to be active above the unit operating between the copper and aluminum tester.

Moving outwards from the cells, spherical shells of torsion were found at specific various distances and intensities. 18 inches and 36 inches were very strong and hot with the tube testing device I designed above.

Joe also lined up all the tubes of the cell and then flipped them over to a state that matched the earth, that is he flipped them to not radiate a heat field. It was discovered that most stainless steel has a cube shaped crystalline structure. This means that, like using quartz crystals on the RainMaker units, there will be a positioning of the material that will effect the torsion interactions of the system. Joe apparently wanted this to be most neutral with the earth.

Distilled Water

On successive charging of various water sources I discovered that distilled water is able to hold a charge much longer then other water sources. It will take much higher voltages without bubbling or breaking down, and the charge will hold for days after the voltage is removed. As well a torsion field is ever present at this state of charge.

It was discovered using this method that my particular cell had one reversed tube. The following are the readings I took after charging the cell, then removing the power supply for 24 hours.

Cell voltages from outer ring inwards

Outer can - 0.0
tube 1 - .3
tube 2 - .62
tube 3 - .35
tube 4 - .76
tube 5 - 1.19
tube 6 - 1.82

You can see that tube # 3 is reversed! Guess I will have to disassemble the tubes, not looking forwards to this one. It will have to be flipped over, and while it is apart I will try my hand at dowsing all the tubes for correct up side to see if this becomes more apparent. [This problem was corrected and now the cell will hold a constant 2.01 volts without ever losing its charge.]

Voltage Effect Observed

This does however show us something very important. The cell is acting similar to a battery, but the voltages present are not the result of electrolytes and plates of dissimilar metals, but the geometry of the cell, causing the charge to hold itself. Probably due to torsional spin of water against SS. Distilled water allows this charge to become greater. There is no explanation for a cell charge of 2.01 volts to remain present on the cell based on any known theory. This charge is present even after trying to discharge it using a short circuit from inner to outer tubes. A battery would be chemically altered by this shorting out and loose its charge. The charge comes right back. Also it is not a capacitive effect but is coming from the water and the cell spin fields interacting. This is the first time I have personally observed a constant voltage resulting from a torsion field

Bismuth Capacitor

The Bismuth torsion capacitor shows the nature of the Joe Cell interaction. It is simply a small Bismuth pellet insulated on one side with electrical tape and placed between two pennies.

The voltages on the pennies from 5 to 60 V dc shows a uniform gradient torsion field is emitted spherically, and the frequency of this field is proportional to the voltage, and strength is proportional to mass of the pellet. This experiment shows us one small part of the Joe Cell, but most importantly we can see that each layer of the Joe Cell must be considered to be one separate torsion generator.

The voltage across the water gap, the thickness of the tube, determine the torsion fields output strength and frequency. To get all the sections of the cell producing the same frequency, all will have to be exactly the same width. This is because the water chains will form perpendicular to the tubes, and altering their length will alter the torsion field produced. The layers of water and SS must be seen as separate torsion generators.

The output of a single layer of the cell is also a function of the mass of the water within it. I have directly sensed the outer layers are far stronger then the inner ones. This is simply due to increased mass of hydrogen atoms with an increased nuclear spin. Frequency is controlled by the gap size. These observations would lead us to the possibility that abandoning the inner smaller tubes and placing bigger ones outside may increase the torsional output of the system. There will be a balance found where the turn becomes weaker and going larger begins to drop the intensity.

Engine Torsion Fields

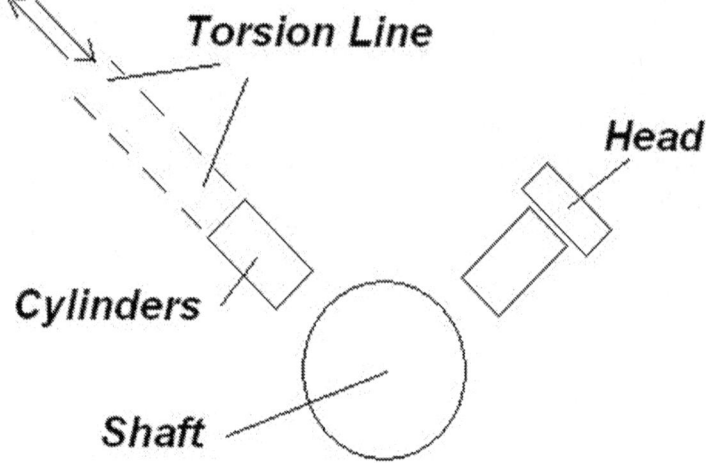

The engine tested was a V6. The Joe Cell was activated setting by the drivers seat, and the presence of the field was highly active and enveloping the area around the car. The torsion lines were located by palming and feeling for torsion fields all around the running engine. It was discovered a linear beam field active radiant above the cylinders where piston motion is operating. The torsion fields are in a state of reversing motion as the pistons move through a sine function of motion, tempic fields are altered for a long distance and radiate well outside the vehicle, probably when the plugs fire.

A scalar pancake coil was held in the torsion lines at a sheer point and discovered a small pressure developing on the coil pushing it away from the piston. Flipping the coil over, the pressure reversed sucking towards the cylinder. Flipping the coil reversed the wind direction from CCW to CW. Indication is that the torsion field produced from the cylinders is not balanced, as spark and ignition happen on only one side of the stroke, this was not unexpected.

A torsion sensor was conceived using a pancake Tesla scalar coil with a 1/4 inch thick Aluminum disc against it. One wire then goes into the vehicle for a scalar sensing device. A touch plate or a scalar coil on a crystal sphere.

It was discovered that the rear left cylinder torsion line passed through the left dash board to the left and forwards of the steering wheel. With the car in motion, extending one hand forwards into the line, and the other down to the Joe Cell I was able to form a mental link flowing through my body. The sensation was a new one indeed, and very exhilarating. There were various motional sensations. The feeling of motion moved into the mind and the body stopped feeling inertial force. The car seemed to surge forwards, without any inertial drag on my body.

Piston Versus Cell Torsion Coupling

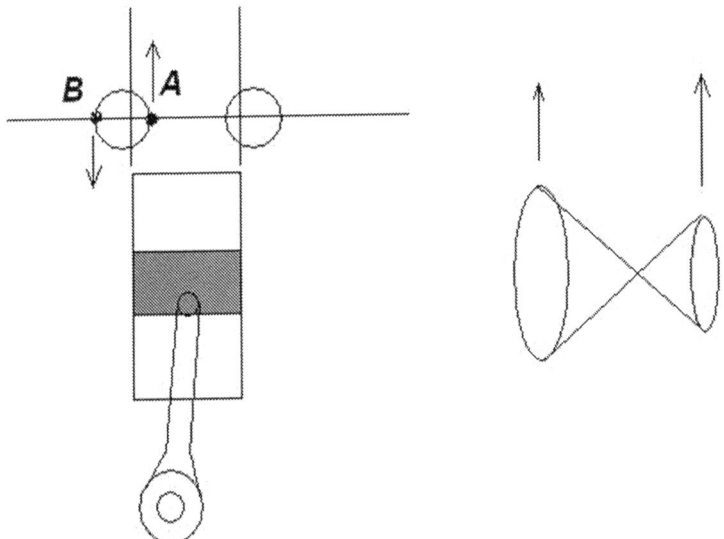

Classic Coupling Method - Torsion Vacuum

An aluminum tube, with a short insulated hose to hold the tube on the carburetor. The process does not rely on any gas or liquid flowing from the cell to the engine.

When the engine becomes equalized with the cell torsionally, then the spark is advanced considerably by moving the distributer timing, and the spark alone creates enough torsion vacuum to set the pistons into motion. Vacuum energy runs cold, and this is a side effect of a density shifting system running on the vacuum side.

With a higher density we have a shrinkage of the physical size of space, while we increase density all motion accelerates, momentum is reduced, and space shrinks. In the piston torsion line I believe that the spark is raising the density of the cylinder chamber and creating the torsion vacuum.

The release of Electric energy happens in steps:

1 - The spark coil magnetic field collapses - a HV pulse heads towards the plugs gap. As the plug charges upwards, E field rises with it and all spin raises velocity at the nucleus of all atoms inside the voltage gradient, where protons and electrons are pushed opposite directions. This happens from the plug outwards by a function of inverse distance squared. The higher density should follow the negative electric potential, but will radiate out much further then the EM field.

2 - The spark jumps - E field collapse - the energy moves into the tempic field, as the magnetic field collapses, and this pushes the tempic field even higher before releasing the energy completely. As all matter pulls back to normal, the area now pressurizes as the piston drops.

Tempic Field Gradient And The Joe Cell:

If you power up only the outer tube of the cell, distilled water and 100 V dc, a smooth field gradient forms and it is impossible for me to locate any sharp standing sheers. The interaction of the water in only one layer of the cell forms a smooth field gradient that falls off linearly with distance. There was no way to make any reading for distance sheers. The spacing on the second tube inwards is approx the same as the case sits from the first tube, just about 1/2 inch. After firing up two of these layers of water still using 100 V dc, I get a very mild sheer forming at 3 feet, but almost imperceptible. The case is thinner then the first tube.

The next spacing is 1/4 inch, and here is where the sheers start to form in the field. The third water layer is being stimulated unevenly with more voltage on metal then water. The metal to water thickness ratio have changed and so the tempic field strength is different and now we get two field gradients interacting off one another, and the fight causes the sheers to form. Going inwards there is one more uneven spaced tube and things get much worse with the whole cell powered. Gradients are following somewhat a 4 - 8 - 16 pattern for some unknown reason, but are sharply present.

The nature of density shifting with the bismuth capacitor is showing no sheers are present, but a smooth gradient forming. Only altering the voltage on the pennies is effecting the level of the field. I believe this observation is critical to understanding the Joe Cell. It shows how each layer is a separate torsion generator and the voltage and spacing sets the frequency.

Conclusions:

Spacing must remain constant on all torsion generators for all the layers to become one coherent field. Water to SS thickness should remain constant.

Since torsion is a linear distance force, the distance of the waters diameter is the critical parameter to the field's strength. Any torsion interactions will spread to fill all matter present. The only way to have all the cells come up together as one coherent field gradient is to have them as identical as possible, in thickness, and distance between them. The torsion generated from a molecule chain of water spinning apparently varies with the distance of the chain. In the cell no water is actually moving, there for the molecules are chaining perpendicular to the tubes. It is the hydrogen's nucleus that is spinning up effecting background density. If there is a spiral gradient effect it is not in physical space, or it is a very smooth one.

The appearance of the torsion sheers, that seem to cause headaches is probably from out of alignment generators emitting different field gradients. And this takes us back to one of our original intuitions. Place a resister network along the tubes to equalize them and get all in resonance. Second choice, separate adjustable power supplies for each water layer. Some more math on water to tube thickness ratios will shed more light, but the distances and thicknesses will probably come up extremely close to identical.

The units in the videos show cells that have gotten water into the burning mode. They appear to favor a 2 inch tube with equal steps up. I have to agree with this construction. To tweak this into a faster torsional rise time, balancing resisters, but in the non powered state the cells will draw off energy, although I don't know if it will discharge them or not. With distilled water and very high pot values we may be able to balance a cell and not discharge it, since if all the layers come into a coherent state the field's power will grow by 4 x each time we double the diameter. Also each cell added to the outer ring will add much more water mass.

This is like spinning up tires on a car hooked to different engines, all to different velocities, and then trying to get the car to run straight. It will cause shimmys and shakes and grind the gears until all the engines get to the same RPM. Each layer of water is one engine. Each can be tuned to the correct RPM by altering the voltage across it.

The answer seems almost too simple. I need to remove the two tubes that are 1/4 spacing pulling the cell out of resonance or set up variable power supplies to each one and tune them manually. Lastly a network of variable resisters may work for tuning the cells.

On Wire Location

On the inner tube it should not matter where you attach the voltage feed wire as to top or bottom, but routing is important. With a DC signal it will be present over the tube at c [light speed] everywhere, and since it is not being varied there is no inductance to factor in. However I recommend inside the inner tube at the top, and better if less wire is submerged as it will shift the torsion as it passes through the water very slightly and disable a certain number of water chains where it passes down through the water on one layer.

Only the water in the center area of the center tube is non active as a generator, and from a geometrical standpoint it should hit the center of the center tube then move straight out to the tube in two directions. Probably not so critical, but it should not come down the outer water layer and then turn back in to the center tube, as it will pass through an area of greater voltage gradient and off balance the cell. This is because one voltage is setting on the case, bring the other voltage within 1/4 inch of this down the water will effect that layer and an uneven gradient will be set up in that water area.

Charging the cell with EM will act exactly like a capacitor with multiple plates. The voltage gradient will move across all the plates in between and the plates between will each charge up based on distance. Voltage needs to appear on the inner and outer plates [or tubes] and no other charged wires should run through the gaps. The metal will short the voltage gradient more and the water will resist it more, so the voltage gradient will vary with the material and is not even.

Lastly the insulators compressed inside the tubes are removing space where water mass can become larger. This is not a problem for layer frequencies, but will lower the strength of the torsion output on each cell. Smaller narrower spacers or even hollowed out Plexiglass sections with very narrow slits and links between them, anything that will make the water lost between tubes more consistent, and lower volume.

Time Effects Noted

The following types of effects were observed by two different experimenters working with Joe Cells independently.

Compelling Observation:

The most compelling observation I have made with the Joe Cell. Active in my new car for about one week, the digital clock gained 10 minutes. I just reset it last Friday. The significance of this has just hit me today.

$E = MC^2$

10 minute gain in one week.

How much of an energy gain this would have resulted in?

How much further would it take to develop 120 Horse Power?

The digital clock on the dashboard is approx 2 feet away from the position of the cell. We can assume a linear drop off of the effect, while the Aluminum tube should be transferring the energy at the cell to the engine.

If we begin to view c as a variable, then calculate how it should effect energy systems, we may come into the ball park with a rational system for applying math to the cells. As well a digital clock can be used to determine cell torsion, and while this may take a week to measure, it is the first method I have found to relate a direct measurement with a device for "non sensitives." Since the background time frame is what we are effecting, it would be very good to explore if reversing the polarity on the cells to spin against earth motion, if this will lower the time flow rate.

At no time was the battery disconnected from the cars clock during this week, the clock has been very accurate for the year I owned it, and this week is proving accurate as well. I was forced to reset it last Friday, as I got into the car to make an appointment, I though I would be late due to the wrong time reading. I reset it using my cell phone for a reference which is synced externally. I kick myself now for not recording it more accurately.

C is [Tempic field force] $E = MC^2$ energy will rise with the square of the c alteration, which will be very small. [10 min in 1 week.]

I come out with [.001 times c] alteration. That is 10,080 minutes per week. 10 / 10,080 ~ .001

Have I raised the velocity of light by a factor of 1 / 1000?, assuming the digital clock is responding to tempic field linearly. However, since we know the crystal in the digital clock is using electric field, it is more likely a factor of the square root of this number and .001 is the actual energy gain happening inside the digital clock at 2 feet from the cell.

It is not important to measure the time it takes light to move between the cell and the clock, for us this is instant for all practical purpose [2 feet]. The important factor is to recognize its "velocity" changing as it moves between them and thus the cell is raising the energy level of the system. From a position where c = {c + (c x .001)} to a position of 2 feet where the cell is setting. The cell must be slightly higher yet. Thus the cells Energy [E] is greater.

2 clocks for a week setting at different distances from a charged cell should give us the measurement for creating a formula including fudge factors and conversion constants if they are necessary.

This is the first tempic field observation I have recorded, and from a conventional background I realize there will be much skepticism.

Dave L

Definitions

Sheer point - Torsion sheer lines - as one palms a torsion field as found in the Joe Cell, as you move outwards are discovered distances where the torsion field is almost as strong as very near the source of the field. It is believed that these hot points in the field are the location of special pressure zones where the platonic or resonant vibrations manifest. Strong fields are layered off the tested Joe Cell at 18 and 36 inches. These sheers indicate that some of the cells are not producing the same output frequency as others and the interaction causes tempic fields grinding against one another. This can also cause headaches at high levels.

SS - Stainless Steel - The Joe Cell is built from concentric layers of only Stainless Steel filled with water between them.

Disclaimer

c_s_s_p group makes no claims as to effects observed, but endeavors to record and share all observations as they happen for general comparison of others work

Dave Lowrance
Dell Coleman
Bernie Heere
c_s_s_p group

Motion

Final release version - contains useful NMR information.

This document is dedicated to the study of motion and mastery of producing motion from the atomic layer and consciousness. The root field force Tempic field, from which consciousness springs as pure vibration. Light is the purest form of motion, and in matter all motion is the result of near light velocity spin.

Knowledge from "The boys up stairs" has continued to be verified, and designs supplied by Kosol Ouch and the Guardians continue to resurface as working concepts. What I have come to call the torsion capacitor, represents both the layers of the Kosol Density travel ship, as well as the layers of the Joe Cell system. This document begins to show the connection between High Voltage pulsing, and Spinning a Mass at High Velocities. Both processes will raise the density or light speed motional constant. NMR charts are introduced as well as some of the other c_s_s_p discussion topics, for those unfamiliar with their meanings.

Further insights on Wilbert Smiths spin concepts applied to light speed, particle physics, and singularities.

[This section is a back peddling effort to reclaim the document after discoveries were realized, it was moved to the front of the document after the discoveries.]

Density

There is a reversal of the terminology when shifting between the conscious and physical layers, and density is one of these reversing words.

In the spiritual pursuits a higher density is conceived as moving closer to the God force, matter becomes higher frequency, c raises, and bodies become ethericly fine, moved by thoughts alone.

In the physical world descriptions of Spin Density we see the opposite, and a higher tempic field density is thought of as the location of slowest c velocity and highest mass center. This reversal of perception will no doubt lead to much confusion, however in the beginning we were not aware of the differences, or where the reversing concepts would be located. Kosol's Density ship is intended to move through higher density from the spiritual perception, not the chemical or physical sense of the word which will invert the meaning.

In this document the word density will be used from the physical perception, and gold has a higher density then water or air. The word refers to heavyness as found in the more dense materials, as well as the Proton having a higher mass then the Electron.

[At present there is still some doubt as to which way higher density lies, in the smaller or in the larger realities, but much evidence points now to the smaller densities.]

Background

Einstein gave us $E = MC^2$

It is easy to see how raising c will effect the energy present inside matter by a squared function. Tesla gave us a concept for faster then light motion with his pancake coil / ball transmission system and recording a wave that travels at 1.5 c. Suffice to point out that in the search for alternate energy sources, any system that will alter the velocity of light will effect all energy systems within the effected area. Wilbert Smith gave us a perception for treating the tempic field as a variable force field with a relative linear distance function. This leads directly to a concept for altering time flow rates and the c velocity that we think of today as a constant. Conscious acceleration is experienced as lightspeed rises and physical density is lowered. As we see with black holes, the time flow rate slows in a stronger gravity field. In a neutron star where the Electron shells "reverse spin" is removed, gravity jumps also. It is where dual opposing spin fields at the atomic level are present gravity is lowered and time flow rate is accelerated.

Affecting The Physical World

As to motion and the physical world we must discern what alterations can effect light speed and which ones cannot, then progress accordingly.

Moving physical objects linearly, is shown to have very little effect on altering c, as a motion of even the earths rotational velocity is very slow compared to light speed.

If we are to effect light speeds, it will be happening at the atomic level where all particles are already spinning at near this velocity.

Bringing two magnets together displays B field force. But pushing the Proton layer of the atom towards the Electron shell, lowering the distance between them on one side is seen to bring two oppositely spinning c velocity vectors closer together. This opposing dual spin system is the very location point for the zero point center of all EM interactions. It can be accomplished with physical rotation [centrifugal force] or with voltage gradient in all the AG materials. The list is now longer to include the components of water.

AG materials [diamagnetic]
Bismuth
Copper
Aluminum
Hydrogen

Magnetic materials [magnetic]
Iron
Cobalt

Nickel
Oxygen

Precession and Pulsing

[The presence of a Magnetic field - aligns spin in the AG metals
[Bismuth, Copper, Aluminum, Hydrogen]]
Electron and Proton particle, spin opposite directions in the
same magnetic field. Orbital and particle spin are aligned so they
follow the precession motions. If we tilt the orbital then the
particle also tilts. While Electron has dual spin Proton does not
and will flip between two alignments. Orbital or particle aiding
alignment. Precession is the result of a torque already identified in
modern physics. The two spin fields move into oppositely
precessing motions where angular interaction is being altered for
the spin fields from each of the other. Maximum light speed effect
is present when there is no precession, raising light speed. As
precession starts, opposing tempic vectors start to angularly skew
from one another and c drops as precession motions accelerate.
Energy moves directly from c into precession momentum.

[It was discovered that the precession of atomic Proton and
Electron shells is a "powered process", that is the precession
motions for the atomic particle shells is the result of a torque. The
torque originates with the inertial momentum of the spin forces.
These are what we term non entropy forces, they do not run
down.]

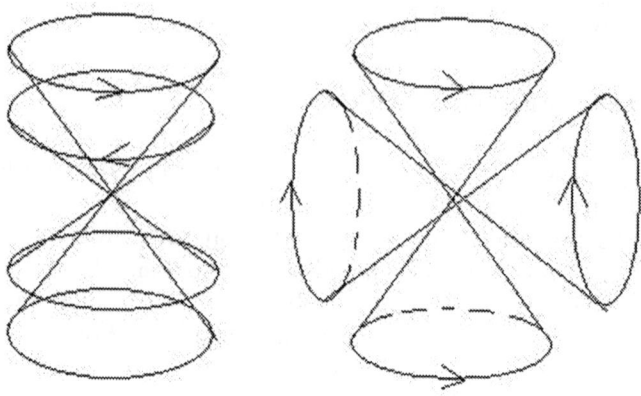

In the case where two field forces precess inside one another opposite directions, if we tilt one spin to 90 degrees we see now the spin directions begin to roll along one another at two opposite corners with an aiding tempic field of motion, and they oppose on the other two corners. Tempic field density now becomes higher along one 45 degree angle and lower along the other. More importantly, electron and proton are spin field decoupled to a great extent and Electron will accelerate while Proton decelerates, both due to internal spin configuration. The atom will radiate a high density field at 90 degrees to a low density field.

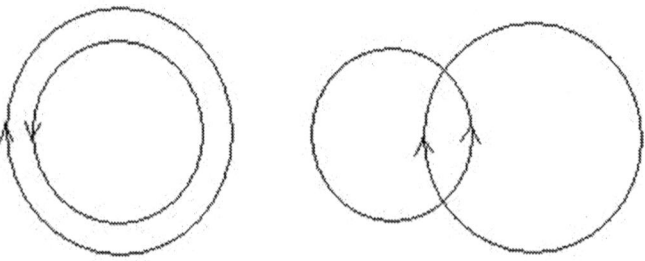

Viewing this from the top we get a sense of the tempic coupling during a tilt. As reversed spin passes the center line the tempic fields fall into alignment and begin to lower c. In the AG materials this couples particle inertial momentum, and alters the torque that creates precession. Magnetic field always pulls the Electron and Proton shells back into alignment, and tempic field always pushes them out. Stronger magnetic field will narrow the precession cones and raise precession frequency. [NMR]

Propagation of Torsion

[Breaking opposing dual spin alignment lowers c velocity in Proton]

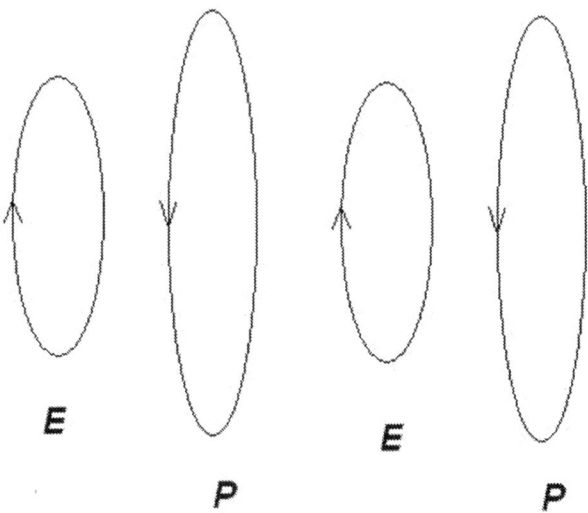

From the spinning cylinder experiments we see how torsion propagates from Electron to Proton, coupling its way along and moving through a magnetic field. Spin is coupled along the B field but it operates at 90 degrees to this as shown in the arrows above. As the particles have actually two conical ends that spin the same direction only one circle is shown for each particle. Within this system of propagation we can identify the dual spin of the earlier models. The most obvious thing that sticks out here is that the Protons torsion component is much larger then the Electrons, while the Electrons magnetic field is much larger. Protons torsion operates "through" Electrons magnetic field.

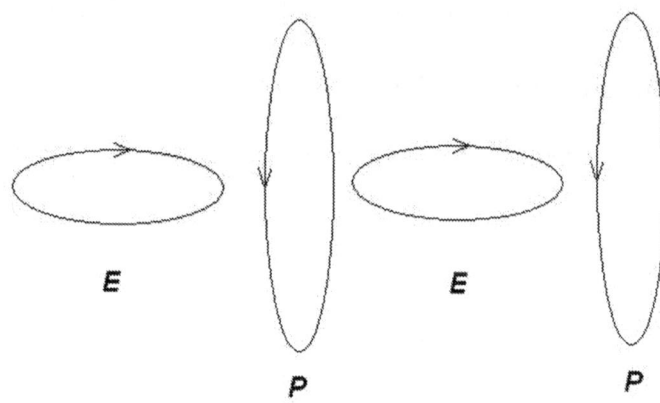

Here we see what may happen if the Electron is quickly hit with an EM field in copper wire. The E field spin tilts first at very high velocity capable of a rotation in the Ghz range. Any pulse that is faster then the NMR rate will separate the two fields for an equalization time as long as one second. A sharper pulse will separate them more fully. Torsion propagation collapses and density increases lowering c at the nucleus. Dual opposing spin disappears at this angle between the particles. Electron accelerates and looses mass, and nuclear mass increases, slowing. Nuclear mass has lost its counter rotating electron influence, and Electrons dual spin can now cancel more fully.

Spin Relativity

Examining spin closely, what happens when a spinning wheel is moved along its spin plane, like the wheel of a car down the street. We find one side of the wheel of the car matches the stationary position of the earth as it touches and the opposite side accelerates to 2x the velocity of the car, where the axle moves to 1X of the cars velocity. Now from inside the car we observe the wheel and both top and bottom appear to be in motion the same velocity [our frame of reference has moved to the axle now]. Observing from the road we see the bottom of the wheel is stationary and the top spinning faster then the axle.

To have motion induced one side of the wheel must have a higher motion velocity vector then the opposite side, and this is the nature of motion with respect to spin. All observers will still agree on what is one revolution, if a white dot is placed on the tire. However all observers will not agree on the velocities that appear along the wheel at the top and bottom of spin.

Consider an atomic particle moving through space with an inertial momentum of spin. Its sides are rotating at near lightspeed.

The act of atomic particle spin moving through space warps the wheels sides and creates a Torsion field. Light velocity spin is covering less distance on one side of the spin, then it covers on the opposite side. What would Einstein say about this? An electron in orbit has a velocity of .999999... 5 times light speed. But if viewed from outside moving at 1/2 light speed, one side of the wheel must be moving at 1 1/2 x light speed and the other at 1/2 light speed in the opposite direction. This skewing with respect to c creates a torsion field. Einstein would say space is warped on opposite sides of spin. If one is riding with the particle however the field is not observable and the torsion field is not present but appears as a smooth spin field with no warping. If one is looking from outside they perceive a torsion field warping every atomic particle.

When any piece of matter is set into motion properties of the matter is altered relative to its internal light speed spin, and time scales start to become altered between what is in motion and what is not.

From outside we perceive an uneven velocity on both sides of the light speed spin field that is in motion. However light speed is perceived as a constant, so what we see is an elongation or contraction of the matter [axle view]. From the particles perception we perceive the world begin to move. No matter what force set the particle into motion, the light velocities will warp on both sides of spin oppositely, and with dual spin they will warp on top of one another opposite directions creating a torsion sheer on both sides of the particles motion vector. Because the interaction is dealing with c rather then a standard linear function, the tempic field is altered. A new relativity model may be derived from this spin model and may add to Einstein's relativity model. Spin relativity.

Earth Relativity

Dell reports on Wilbert Smith research,

"Inside a spin field everything is relative as Smith points out. But since we are always inside some kind of spin field
(planet, solar system, galaxy) there is always a clock that is the correct clock to use - On the planet it is earth rotation, in the solar system it is the sun,
going to Alpha Proxima it is the galaxy. By having an absolute framework (if you pick the right one) we always avoid the Einstein paradox.
If you are seeing differences in spin direction from different positions then you don't have the right clock and need to move out one spin frame (at least)."

When designing any torsion device, earth spin must be considered. Generating a field that predominantly adds CW or CCW mass spin to the nuclear area of the atom, one direction will add to the earths spin field and the other will first subtract from it, before increasing again in the opposite direction. If we are to navigate the earth, all will be relative to this gradient. If we are to navigate the galaxy then another spin gradient will become our base. Turning the system upside down will reverse the dominant spin, and moving between hemispheres will turn it upside down with respect to earth. The space program always shoots rockets off towards the east so that far less energy is required to reach orbit. The same is true for the Joe Cell, and to get the same power from a spin effect, turning the correct direction is essential to using the earth's energy to help power the device. If your Joe Cell is reversing the earths natural spin direction then it will have to work much harder to get the same results.

Utron Model:
[Drag Causing Motion]

If we examine Wilbert Smiths basic concept of "spin" or "light speed spin". We come to see all matter as dual oppositely spinning masses powered by light, the balance point being the Zero point for this density when motionless in space. Now we examine Otis Carr's Utron model.

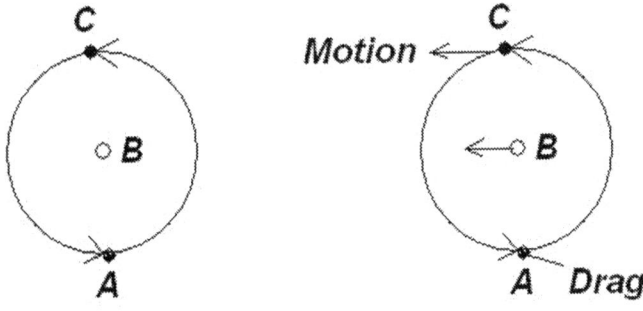

Applying drag to spinning copper cones creates motion, and the harder we try to stop the spinning cones the faster they shoot off into motion. As a pure tempic field the motion is along the spin plane.

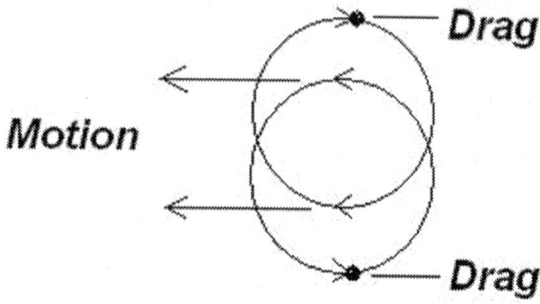

Dual spin results in linear motion if drag is added to opposite sides of the dual spin fields, and the two spin fields are coupled along one axle.

[Physical spin raises c].

We see from the above diagrams that if we tried to stop both spins from the outer sides, the result would be the inner spin would accelerate and the system would move into motion of the inner arrows.

With two oppositely spinning masses of Copper, a diamagnetic material, placing a dragging magnetic pulse on opposite sides of each causes the spin to induce motion from both sides equally. As the spinning cones are hit by the magnetic field, it is accelerated away from both spinning cones the same direction. This is one model for motion from a tempic field. It is motion produced from adding drag at the correct places of dual spin.

In the Utron, we see two spinning masses being pulsed on opposite sides at 45 degrees. What is not obvious is why this would propel the system faster then the motors are spinning the cones. The Coppers Protons shell will be locked into the physical spin field and pushed outwards by centrifugal forces, shoved into the electron shells that bond the cylinder together. Two light speed velocities moving opposite directions are now reduced in distance. When the magnetic pulse hits the cones at high velocity the electrons magnetic field will tilt to 45 degrees and one magnetic pole of the atom will turn into the side of the atom being compressed, and the other will turn into the side being expanded. The nuclear spin will not turn as fast, and it will be now setting in opposition in each cone while the electron spin effect has been removed. Proton spin will meet proton spin from each cone directly with far less electron involvement. Protons will be cancelling their tempic field density, while Electrons are also free to cancel their own tempic field density.

Bismuth Torsion Capacitor

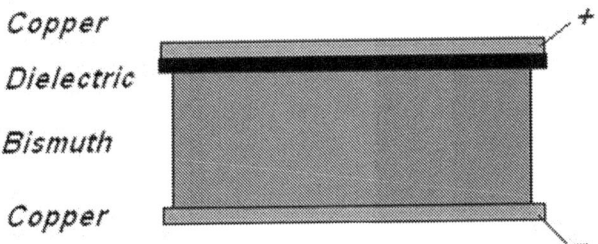

Voltage Gradient - Tempic field distortion in the AG metals

[The presence of an Electric field - raises c on both magnetic poles of spin equally - Electric field effects both Proton and Electron equally in opposite directions]

In the Bismuth Torsion capacitor, placing a layer of bismuth insulated on one side between two layers of Copper we see and feel a tempic field interaction when charging the plates.

83

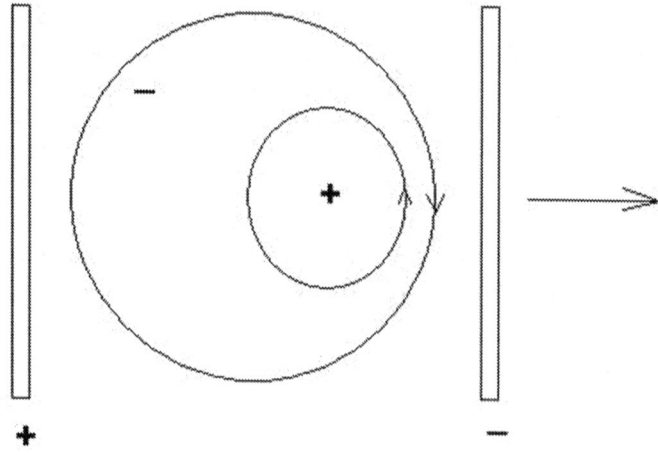

Proton shell is pulled towards the negative plate and Electron shell is pulled towards the positive plate. The Bismuth atom undergoes a light velocity alteration on each side of its orbital spin circles similar to the one above with motion. The area to the right, the opposing tempic field lines counter and accelerate c, while on the left side they move away and lower c. There are also precession angles to consider but over all it would appear that the atom would tend to move upwards in c velocity if this effect was a strong enough one. The torsional distortion from light speed moving slower on the left and faster on the right is the same sort of distortion we see with atoms in motion through space approaching light speed velocities. As tempic field density lowers, light velocity increases and the effective distance light covers becomes larger. The atom increases size. If the voltage is quickly dropped from a fast discharge, then a physical vacuum may result.

If a magnetic field is present and used to align the Protons spin plane such that one pole moves into the compression side and the other pole moves into the low density side, then we end up with an atom having one magnetic pole in a higher density then the other. The precession cones of the magnetic field are now distorted and one side takes on more energy. If pushed far enough we may see a monopole field appear to emerge.

Tempic field interacting inside circles are a function of perimeter distance sums, as they are a linear distance force. When centered exactly perfectly the atom will be at total sum lowest density. This is because any movement off center decreases the top and bottom distance sums. While the sum of distances on each side remains constant, the top and bottom sum becomes lower, and so there is more overall tempic interaction introduced between the Electron and Proton shells by one another. Tempic fields acting to oppose one another accelerate the light speed velocity, and all atomic spin is at near light speeds.

The AG metals can produce this effect, but then they will radiate to any matter around them directly nucleus to nucleus once they become a torsion field.

In the Bismuth capacitor once the higher torsion is generated it seems to permeate in all directions equally. In the Joe Cell systems engine water is where we expect to see the highest effect generated as its hydrogen atom will be subject to this effect when the plugs fire. If the charge is building on the plug at the top of the cylinder, as it arcs, a higher density gradient may form at the top of the cylinder creating a torsion vacuum.

Field Forces And Atomic Interactions

1 -The presence of a Magnetic field - aligns nuclear spin in the AG metals [Bismuth, Copper, Aluminum, Hydrogen]

2 - The presence of an Electric field - raises c on both magnetic poles of nuclear spin equally - Electric field effects both Proton and Electron equally in opposite directions.

3 - The presence of high velocity spin - raises density on both sides of spin equally - Protons are pushed outwards as Electron bonds hold the matter together and pull back against them.

4 - Magnetic field effects Electron shell strongly [Oxygen and Iron] and Proton shell weakly [AG metals].

5 - Torsion field effects Proton shell strongly and Electron shell weakly

6 - A magnetic field in line with an electric field will raise c and make one magnetic pole become dominant [Joe Cell]

7 - A spinning mass [raising density from centrifugal force], at 45 degrees to a pulsing magnetic field will pull one magnetic pole into higher c velocity to become dominant [Utron]

Joe Cell

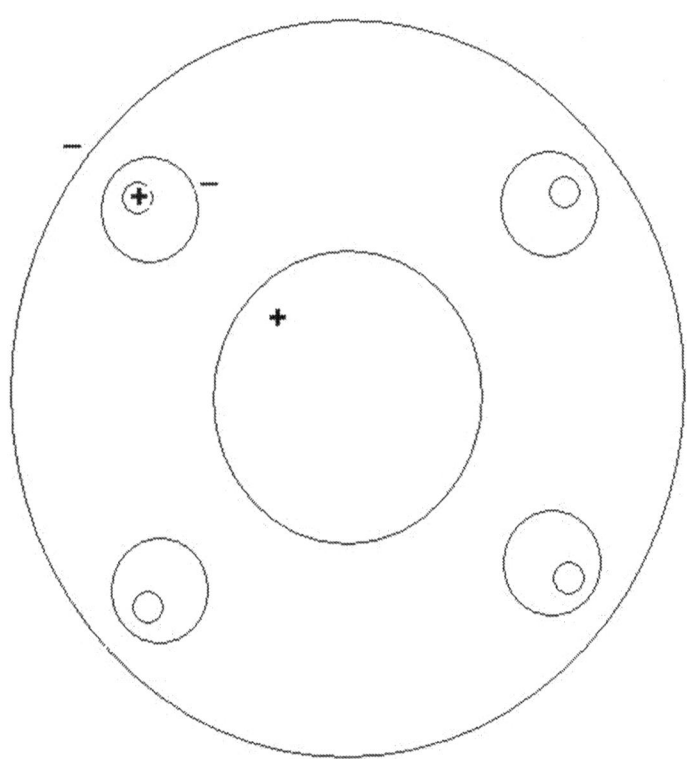

Initial Charge

This diagram illustrates the voltage scalars present on one Joe Cell water layer, as a circular charged system. This is the same result we would see if the circle was set into a spinning motion, nuclear mass at the proton layer would be propelled outwards

87

pushing it into the electron shells that hold the bonds of the device together. The E field is acting like device spin would. You can see with a Northern Hemisphere charge on it, the Hydrogen atoms nucleus receives an outwards push evenly all the way around the cell. Now if the magnetic field turns outwards one pole along the outer ring and one pole towards the inner ring, one side of the magnetic field gets a higher torsion state then the other inside all the Hydrogen atoms, as they will align with a magnetic field. The one being pushed into the electron shell moves higher in spin and the other moves lower. The dominant magnetic pole facing outwards around the tubes.

Water Molecule Chains

The molecule chains of water will break open in the Joe Cell and attach one side to each wall of the tubes, lining up perpendicular to the tubes and forming chains across the tubes gaps. The ends touching the tubes will be free to vibrate into circular motions, as the charge along the tube is even, not so with the other water molecules that will attach end to end. The nuclear spin motion starting along the water to SS surface will now propagate along the molecule chain and start to move the Oxygen atoms into a circle setting up a magnetic field in line with the voltage gradient. The nucleus of the Hydrogen atoms will be pulled one pole into the electron shell and one pole free of it. This will imbalance the magnetic field. Since all the molecule chains are the same length, the torsion motions will become coherent like a whole bunch of little clocks next to one another the size of one molecule. It is the distance between the tubes that will determine the properties of this molecule chain, mainly length and voltage across it.

Piston Torsion Line

Our Joe Cell research has shown us that the pistons of an internal combustion engine generate a longitudinal wave front that leaves the engine in the same direction as the pistons motion vector. This field can be sensed as a torsion field and it represents Tesla's longitudinal wave which is also produced by rapid pulsing of EM fields. It is most likely the result of the E field pulses happening inside the combustion chamber, as physical motion is not fast enough to warp light speed motions, however we see from the above that E field pulsing will if it hits a diamagnetic substance like water or Aluminum. And motion will also do this at the point of reversal, but in sine wave motions, like the pistons use, the reversal is gentle. The action on the atoms however is the same, both the reversing motion of the pistons and the electric field will push the atoms into the same unbalanced state with atoms mass [nuclear mass] being shoved strongly into the electron shell on only one side.

Cell Models

Several models are suggest at this point of Joe Cell experiment, and need to be fully explored.

Lazar Beam Model:

Similar to a coherent light beam, or Lazar, if all the Hydrogen atoms in the entire Joe Cell at all layers begin to spin along one spin plane, as well move at one common frequency, then they will begin to form a coherent torsion field who's power level will grow expotientially. This is done one atom at a time. As the water molecules each begin to align with the charging voltage they will line up slowly such that all the voltage charges begin to stack like little batteries. If they all remain locked into this orientation then the cell will output a natural voltage after charging is complete. Now the goal is to get all the cells layers producing torsion at the same frequency. The outer layers will have a higher mass so will be stronger, but the frequency must match. This is a function of the distance between the tubes and the voltage applied. Tuning resisters have been suggested to balance layers of the cell with uneven distances, and smaller water thickness will need its voltage reduced to match frequency. This is done by placing resisters across it until resonance is felt and all torsion sheer lines disappear. [This is questionable at present] If this method is correct then all the torsion sheers should vanish to a smooth gradient.

In this model all the water receives the same spin charge field and becomes charged equally.

The root concepts here are "coherent tempic field", and "coherent light" which have been identified as the same concept through tempic field theory.

Rising c Velocity Model:

In this model the c velocity of the cell rises towards the center, and it may be expected that lower spacing between tubes would create multiple torsion sheer layers towards the center of the cell, where the highest c velocity would appear. This is based on some function applied to alter the density by steps. The result is that the water in the cell would now receive multiple layers of spin charge. They would not become coherent but each push off the next lower one to reach a higher state at the center, much like a tornado with tighter spin towards the center. Torsion sheers will form outside the cell where the multiple fields interact and move along one another at different spin rates. Reverse patterns of spin will form between the layers moving at different speeds setting up platonic forms.

While it was discovered that reducing the cell spacing by half may cause a reverse voltage to appear in the cell, indicating a reversed spin field forming between coherent layers, if this was done more gradually it may work and hold its charge. The cell would have a higher spin field charge at the center layer, and a lower one along the outside. The reverse of this model is also possible where the outer layer becomes a higher frequency, and due to the increased mass of water present on the outer layers, smaller spacing at this end would be more effective.

Foot Note On The Light Body Meditations

Remember for those who do meditations around this device with light body methods here is the method again in a simpler form.

1. Position lying down, sitting on a chair or you can sit in full lotus, half lotus or buhda posture for meditations.

2. Close your eye, visualized a center of the head sun in the center of your brain. If you visualize a line from top of one ear to the other top of the ears. Then you visualize a line from the center eye or sixth chakras to the back of your head, and where these lines intersect that is where you visualize the center of the head sun also it is where the hyperthymus glands is which contains the pineal glands and pituitary glands and the visualized sun looks like a sun small as the golf ball size with it is brilliant and radiates light shining just like the real sun but it is small size like the golf ball.

3. Now you visualized the sun in the center of the brain and you breathe through your nose then say in mentally at the same time you visualize the sun radiating inside the center of your head. Then when you breathe out through your nose you say out mentally and at the same time you continue to visualize the sun in the center of your brain. You repeat the process over and over until your breathing become shallow and so soft that even you begin to realize that you can breath without breathing, you are getting so relaxed, you can continue to be in this stage of meditations as long as you want but continue to repeat the above practice of slow, shallow breathing and center of head sun visualizations. No limit whatsoever.

Platonic Vibrations of Concentric Tubes:

The spherical harmonics of the pipe diameter should be present in every tube at different frequencies, based on tube diameters. This is a vibrational field moving around the tubes as filled with water, and as the vibrations reach a certain frequency we get a resonance in one of the layers of the Joe Cell. If all the layers can be vibrated together we may see something happen. Once again either separate power supplies to each tube for experiment, or a series of potentiometers to slowly adjust the balance until all layers of the cell are in a state of resonant vibrations. One can energize one layer at a time and record its resonant peak voltages, then build a resister network from there.

It was observed early on that altering the voltage across the Joe Cell produced several resonant peaks, it is possible we were hitting one layers resonant peaks at a time, and over the spread we see all of them. It was also observed that these resonant peaks are not present in the bismuth torsion capacitor indicating it is probably coming from the tubes circular shape. [Ongoing experiment is in progress to more accurately describe a usable model for workable designs].

Experiment has verified that vibrational patterns of a platonic form only manifest if the cells have different spacing, and the patterns forming outside the cells are stationary interference patterns.

This indicates two different torsion fields interacting off one another. In any cell with only one layer of water energized there are no platonic vibrations that I could find anywhere, and torsion rises smoothly with voltage gradient.

The Roll Of NMR Charting In Alternate Energy Work

"The roots of motion lie within the atom, and at the nucleus is the heart of the major mass which is in a state of torsional light velocity spin.

I am amazed at the current desire to define all EM interactions without going any deeper then the electron shell. In truth very few of the elements respond to EM usefully at the electron shell and most are magnetically neutral. It is the nuclear shell that more often responds to EM uniquely and alters the atoms strongest field force, that of spin, or tempic field." Dave L

Background

"Use the online NMR Calculator. Go to the top where the chart can be programmed for magnetic field strength. Plug in the value for hydrogen at 1 Tesla which is 42.58 Mhz/Tesla. In the box marked 1H frequency enter "42.58" hit enter. Now go down the chart and you can read the frequency for any isotope of any element setting under a neo magnet as we use in the RainMaker. This will give a close approximation for tuning the function generator. As neos range from 1 T to about 1.5 T it will be close enough to show you what function generator you will need to get to hit NMR rates. In any one device with magnets you will find resonant layers moving away from the magnets that drop frequency, but many peaks will be sensed as you tune through the band. They are sensed in the mind just as scalar coils can be."
Dave L

A Chart With Highlights

When deciding on elements on the chart, two things are very important. Abundance and magnetic moment. Look at Bismuth for the clues 100 percent abundance means that every single atom of Bismuth is consistently useful. Same with Aluminum. Look at copper and you see it is split into two isotopes, each having slightly different frequency but both adding to almost 100 percent useful atoms. Look at iron, only 2 percent abundance, very few atoms of iron will respond to EM interactions at the nucleus. Iron will be very hard to align the spin to achieve a coherent field of any sort using magnets. While iron will still follow a density gradient, it will not generate one of itself. Iron is useful on the electron shell where it is magnetic.

Abundance

Abundance is the percent of all the elements atoms that have the isotope on the chart. Some elements only have 10 percent abundance that will respond to NMR. We assume the rest are also present with other isotope configurations because in nature they usually always are. If they scan an MRI and find a certain quantity of magnesium, they must increase the reading to reflect the other 90 percent that will be invisible to the MRI scanner. They will not be detected because their isotopes are different. Abundance is a best guess and comes out accurate enough for internal pictures to reflect what is actually there, and not merely what is detectable. There are a multitude of isotopes that are not on the chart, these others are not on the chart because they do not respond magnetically. We cannot align them using EM, we cannot flip them over to a high energy state, and they will not produce photon releases we can detect. 90 percent of all magnesium atoms cannot be manipulated using EM but they will be found to be present if the sample came from the earth naturally. These counts of atoms are extremely high. If they did not add in the calculation for abundance then a sphere of magnesium inside your body would appear on the scan as 1/10 the size you would find during the operation. Abundance is on the chart for a very good reason. Bismuth is 100 percent abundance, this does not mean it is the most abundant element on the earth, I believe that Aluminum is. It means that if you go out and mine Bismuth or Aluminum, every atom in its mass will respond to EM manipulations, and absolutely none of its nuclear isotopes will be hiding from it. All will produce the same photon frequency in the same conditions.

Nuclear Magnetic Moment

Magnetic moment will determine which direction non canceling spin will be turning, in a magnetic field. While these magnetic fields are very small and do not do much outside the atom, the spin attached to them is the strongest of the atom, and nearly all the mass of the atom is located here at the nucleus. Nuclear magnetic moment gives us the overall direction that the nucleus will spin in an external magnetic field. You find positive and negative numbers in this column.

Examining water we see O and H have opposing spin fields, and O is not very useful as its abundance is very low. Adding Chromium to the mix raises the abundance to 9.5 percent with a like spin as Oxygen. Chromium from the Stainless Steel will dissolve in water and is not dropped out by the voltage on the Joe Cell. This places a mass inside the water that has opposing spin to the dominant hydrogen spin field. It will align if a magnetic field is present. Hydrogen is almost 100 percent abundance and will dominate the system. Hydrogen is the reference element for all NMR work because it is most identifiable in MRI scanners and easy to spot.

This chart and interpreting spin fields gives a new light to all the magnetricity devices using other metal compositions to create rotating motors and devices with magnets. Since these fields are torsion and tempic by nature, it is the nuclear spin fields that will determine how the systems operate inside the magnetic fields present.

While this chart is used for precise manipulation and detection in very strong magnetic fields that are very consistent using super cooled magnets, we can still use the information to determine spin direction as well as general frequency for setting pulsing systems that manipulate torsion. A pulsing system however alters magnetic field and sweeps a band of useful frequency photons. For atoms to absorb energy, a changing magnetic field will spread the frequency out that can be absorbed by the materials.

Useful Interpretations Of NMR Data

We can use the NMR charts to locate the dream elements like Bismuth, and also to determine which direction its torsional spin will be turning if an EM field is present. This is very necessary to know when setting up a nuclear torsion powering system on the earth. In past times we have always "dissected" things to study them. With atoms this does not work. Smashing them and blowing them apart will not teach us how to pull out nuclear energy without destruction of the atoms themselves. The necessary concept to be grasped here is that nuclear spin is a force of nature as the atom stands, in its complete wholeness, and the atom will always pull back to c velocity spin no matter how we use it. This comes with an understanding of the tempic field, torsion, or spin fields. Blowing the atoms apart we observe a very many short lived energy spin pieces that then disappear, and we label them sub particles, yet maybe all we have really done was create them in the destruction of the stable particles.

Nuclear Spin - A Deeper Look At The Past Knowledge

Andrew reports some current research,

The Baker patent references the Wallace patents. It also contains Wallace's "little known heat pump patent" that Ron Akita referenced in that correspondence I had with him. The heat pump patent, utilizing nucleon spin polarization, is Wallace, US3823570

From the summary:

In the past century great strides have been made in harnessing the three degrees of translational movement of electrons in the electromagnetic regime. Very little, if anything, has been done to utilize the inertial regime comprising the relatively massive nucleons, and particularly the three degrees of rotational freedom thereof one of which is preempted by nuclear spin.

In my U.S. Pat. No. 3,626,606, I demonstrated how, by applying a rotational force to material whose nuclear spin number (I) is half-integral, a reorientation can be achieved in the nuclear structure. In my U.S. Pat. 3,626,605, I demonstrated how a time variant may be imposed on the output resulting from such reorientation.

It immediately becomes evident that there are potentially many, many uses of these reoriented nucleons. In all probability, techniques will be found which to some extent parallel those employed for utilization of the electron in the electromagnetic field, and it becomes clear that the thus reoriented nuclear structure may lend itself to such uses as modification of the gravitational field acting on a body so as to alter its gravitational attraction toward another body, separation of isotopes by distinguishing between nuclei according to their nucleon content, generating of gravity waves for communication and other energy transfer, stabilizing of plasma and maintenance of plasma density for controlled nuclear fusion, possible harnessing of cosmic gravitational energy in addition to utilization in many, many other fields.

Dave, Wallace Adds This To Your NMR Discussion And List:

The Spin Values (I) for the isotopic forms of the elements are well known and may be found in tabular form, for example, "The NMR Table, Fifth Edition," published by Varian Associates of Palo Alto, Cal. Typical among substances in elemental form having such half integral spin values are beryllium with a neutron spin of $I = 3/2$, aluminum with a proton spin of $I = 5/2$, chlorine of which both isotopes provide proton spins of $I = 3/2$, vanadium (useful for alloys) of which one isotope of 99.76 percent abundance provides a proton spin of $I = 7/2$, cobalt with a proton spin of $I = 7/2$, copper of which both isotopes provide spins of $I = 3/2$, and bromine of which both isotopes provide proton spins of $I = 3/2$. These chemical elements and others of half integral spin values may be alloyed together as well as to chemical elements possessing no-spin and integral spin nuclei provided quantity percentages of such additional elements are small.

Conclusions

Wallace had already studied this NMR table over 50 years ago and determined its use in nuclear energy as a mass spin field effect. He spun up these elements to very high RPM and studied the field force interactions, even hitting on gravity. He directly relates this to the nuclear spin, and inertial momentum of the nucleus of the atoms.

What we need to realize from all this is that all the elements on the NMR charts are torsion field generating materials if EM is present, and all will contribute to a torsion field in some way if present at the point of generation of the fields. Some will add positive spin and some will add reverse spin, and we can look at the charts to determine this when designing a system.

Vocabulary:

Density - Layers of the universe where light speed appears to be relatively constant, therefore pushing density is altering light speed [Also see description above in the text.]

In chemistry, the density of a mass indicates its weight compared to an equal volume of water.

In Spirituality higher density is moving towards the God force [Source].

c - abbreviation for light speed, thought to be a constant in modern physics, but offered as a variable in the works of Wilbert Smith

Torsion - as dual opposing light velocity spin is thrown into an off centered state, torsion fields are sensed from the physical side - as heavy wheels are put into rotational motion centrifugal forces create torsion by hurling the nuclear level outwards into the electron layer. Torsion is referenced from the physical side and tempic field from the c velocity side.

Tempic field - represents light velocity relative motion and all atoms particles have inertial momentum vectors relative to c that drops towards the center of spin geometrically. The field drops off linearly in space - opposing vectors raise c and parallel vectors lower c on the physical side, this works opposite on the conscious side

B field - In electronics the magnetic field is given the letter B, this is also referred to as the M or magnetic field in many of our local references and is one of the T E M group of field forces.

T E M - The field forces defined by Wilbert Smith, Tempic, Electric, Magnetic

Contributing Authors:
David Lowrance
Dell Coleman
Andrew Bellon

A New Focus

6/3/2007 David Lowrance [c_s_s_p group]

Rainmaker Activation:

Today I am opening a new CU. This is a sacred process of bringing a new "life form" here to share and overlap my living space in the physical. I have dedicated this to the physical mastery of motion, and to aid the comprehension of the mastery of motion in the physical world of our density. The tapping of the prime force of nature directly.

Thoughts flowing in are the prime reality for a conscious being and the evolution of consciousness involving the following connections. Each is an opening.

First the individual, and self awareness, I must care for myself and the state of life force within me. Extension of the life force outwards to all others indicates a mastery of myself is present.

Second the connection to the consciousness of matter that shares our physical realm, and in particular, my personal connection with the focus of the matter in my immediate area, or living space. I have asked the CU to be present in my boat, and the full area of the boat.

Third the matter of the earth, which comprises a consciousness of mother earth. My boat lies within a sacred line of the earth, the line moves through center from stern to bow almost dead center. The odds of this happening randomly, in a stall that I would not have chosen are incredible, but from a higher perception not unexpected. This gives me a unique access to mother earth, and must have been part of the sacred plan on some level. Also the presence of salt water is viewed as something possible necessary, and I will not limit myself.

Fourth the collective consciousness of mankind and the brotherhood of our evolution to a higher place where the shift is welcomed that will free us on the astral plane, and release our mental process to freely soar. As we are still in a current war, this is only a hope for mankind at this point and something I will continue to observe as others of our planet choose to grow and seek the higher perception.

Each step is now viewed as a sacred process, and any such new CU must overlap all the levels possible. The bigger picture must be present along with the minute details of the physical in the endeavor to harness the wheelwork of physical motion, if it is to become mainstream. The higher connection is present, but the goal is not one of transcendence. At this point it has become a given, not to become lost in the Light, but to use it as needed. It is not transcendence of the physical realm that will alter mankind, but the mastery of our present location with respect to time, space, and motion. The beings of Light will not do this for us, and the slow struggle to achieve comprehension is a physical mastery that must reach down into our physical minds. This has always been my goal from the start of my alternate energy work.

Development Of The Second Opening:

Where matter is located, also will be found focus or focused awareness, it is this awareness that provides the force behind physical motion. This force is the prime "Source", the tempic field as we observe it. Torsion or torque is the result of any such release. The presence of mass and alteration of focus of mass to a single intention or physical point will amplify consciousness itself.

Power

To a society looking to physical devices as source of energy, power lines for transmission, and fossil fuels for vehicles, the world is a fixed material manifestation with only entropy powering it. We simply release what we observe God has created, and then we leave the garbage to be recycled by the God force, with little observance of the looping and spiraling nature of the process. It is only in the magic of Lightspeed velocities that we find ourselves lost in wonderment. A force is present we can not explain with our current models of physics. Also this force is not limited by astral interactions, but may be focused by choice from both astral and mental realms in concert. The only force present that appears to be inexhaustible, that of Light motion.

This leads to the conclusion that to advance we must master our emotional existence first, and a state of joy and wonderment must be constantly present. Sending the Light to all other beings becomes a given in such a society. When the Love begins to be returned by all participants of our collective mankind then the whole will raise in frequency. Power can only be controlled from a higher level. The same will be found true in our physical realm and the military secrecy can only be released from a higher perception. Unity is found at the velocity of the Light, where we truly exist as Lightbeings. All beings will realize that hurting any one individual hurts us all as a collective.

I am now fully convinced that motion can be induced into any substance, any material, and the process will envelope any and all matter inside the immediate operating area of a CU.

A density shift is possible in any material. If the consciousness is strong enough it can manipulate the physical space location of anything. However we are not at this level, and we must rely on the alignment of materials we can "hook" into using the lower forces or EM. The remaining T field alignments can then be directly harnessed by a weaker mind mass, or brain mass such as humans currently have present. Secondly a linking of many minds, 5 or more can achieve this directly but with the alignment of physical objects the one can also do it.

Two Methods Observed:

Instant repositioning of matter in the physical, using the conscious directly [Teleportation]. Only the conscious projector needs to be re focused relative to the background Aether. This process involves first releasing or crashing the torsion field in matter at the physical layer along an entire envelope. Devices can be created to aid this process. The increase of conscious focus by harnessing matter to add its focusing or power is necessary to achieve this. In the physical we can manipulate the forces of EM to align the conscious projection into a state of unity, then steer the resulting connection from our own conscious layer. [Carr]

Physical motion can be induced into a mass. [Joe Cell system]
Gravity can be disconnected directly - party levitation

The methods all use the conscious layer at some level, so this is prime mastery, or the first step. Process and conditioning of the human is the first goal. At present we see a random theology being used, but all lead to the common truth no matter the specific vocabulary imposed. It is only the limits that must be identified and expanded by a comparison of all that truly works. The perception that all mankind is one is essential.

In the operation of Light we find the miracle that is ever present. In the manifestation of motion and the interactive forces of EM we discover the dance. This is a triune manifestation, and the balance is therefore not a linear system, but one inducing precession of spheres. To achieve linear motion, we will have to harness the movement of spheres or cylinders. At the atomic level of source we have spheres with field densities that are not perfect but fall into areas of operation that can be separated.

Motion in spheres, and altering field densities of the field forces in spheres [atoms].

Conscious connection, and the ability to alter spin or incident angle of the field forces mentally.

These are the keys I can identify at present to produce motion using only Light speed spin.

Magnetism can be used to align either electron or proton shell spin in the material that respond to these "hooks".

Electric discharges can be used to align torsional spin as well that will propagate a mass.

Torsion, the background field that connects all matter through a network of connecting motional energies between nucleus of all atoms present and also the conduit of consciousness.

The force of the "higher mind" acting on space to create matter.

Postulate

There are two types of electrical arcs:

Excess electrons [negative charge] on the outer shells of electric materials, will arc to a less electron populated material. The forces that become radiant or dominant are electric E field, and nearly all the nuclear hooks are disabled, due to the atoms local balance cancellation of charge. Electron shell is dominant in this type of arc.

Absence of electrons [positive charge] will draw or suck electrons from another material object, due to the atoms extension of the positive charge of the Proton shell.

Dominant force present in the E field is now originating within the nucleus of the atom and the strong force area.

Each of these field forces are E fields of equal potential but only one is "hooked" into the nucleus of the atom. In atoms of magnetic hooking as with the AG metals, positive arcing will also align the Proton shell, but there will also be a 90 degree shift or inductive interaction.

In atoms like iron where there is little torsional interaction between electron shell and nucleus, positive electric arcs will tend to tilt the nuclear mass into alignment with the path or direction of the arc. In atoms with inductive qualities the nuclear spin will shift to 90 degrees of this angle.

With negative arcs, on iron we see little to no interaction on the nuclear level. Thus a positive charge on iron is seen to manipulate nuclear alignment, but not a negative charge.

On Aluminum we see an interaction with both types of arcing, however it is a complex interaction involving both electron and proton shells.

Using a scalar coil we can separate the Proton and Electron E field density areas. Then placing a positive charge on the iron we can then manipulate its inertial spin angles using arcs to a less positive charged material.

Using both iron and Aluminum two fields can be manipulated in the iron, E and T fields.

Experiment:

An experiment can be set up to study the difference in the density response of the metals. And the basic goal is to produce motion in the materials studied.

A scalar coil with iron core charged positively. Now arcing is set up using an automotive system coil, or multiple coils to study the angular interaction, and how the iron is effected.

To study the interaction between different materials, layered coils can be present as well but arcing must be controlled in angle for each metal present.

Note the effects to density, and if an angular torsion can be achieved, that is, more torsion concentration or aligned in one direction. Motion will be the result if an angular torsion field is present at a high enough level. The material set into motion would be expected to be the positively charged metal.

The ideal test setup would supply 12 auto ignition coils and means to pulse or fire them all at once, or in succession of an ordered pattern of frequency.

NMR tilts can last up to one second, and commonly 1/4 second so frequency does not have to be high, as the mass of iron is quite high.

Foot Note 2.

Just remember the light body method can be used to facilitate stargate travel also as well this meditations can be done in close proximity of the RainMaker device, the consciousness ship as well the Joe Cell device. No limit whatsoever.

Also as for the Joe Cell system device, the inner cylinder can be filled with salt water, the middle cylinder can be filled with vinegar and the outer cylinder can be filled with fresh water, the vinegar and salt has opposing ph level, the salt water has negative ph so that will create inflow when in contact with torsional fields from the activations of Joe Cell or the Joe Cell put over the RainMaker device.

And the same is true if the vinegar is positive ph level so that will create a outflow energy when introduced to the torsional fields generated by the Joe Cell activations or the Joe Cell unit put onto a RainMaker base units.

Now the cylinder of the Joe Cell is sealed from each other by the hot glue from the glue stick and glue gun so there is no leakage from cylinder to cylinder.

The fresh water section of the out cylinder will create the diamagnetic fields effect for the inflow and out flow energy from the vinegar and salt water once introduced to the torsional fields.

This fresh water will create diamagnetic ionizations around the Joe Cell and the RainMaker base.

The RainMaker base can be increased it torsional fields by introducing the torsional capacitors into its base. This as well will increase the Joe Cell torsional fields also, now there is many ways to take power from the Joe Cell system once you set it up according to the above set up descriptions, you can connect two leads, one for the vinegar cylinder while the other is for the salt water cylinder in the Joe Cell. As you may know by now, this two different cylinder in the Joe Cell contain two different liquids one is salt water is a base ph level contents that is inflow when torsional fields is introduced to it, and the vinegar is acid ph base content that will be outflow or hot flow energy when the torsional fields is introduce to it by the activations of the Joe Cell or the RainMaker device base. Both the Joe Cell and RainMaker device has an outflow hot energy electrical layer and a cold inflow electrical layer. Inflow is clockwise spin of energy while the counterclockwise spin is the outflow layer. There are many layers of this type of torsional flow in both the Joe Cell and the RainMaker device, so there is no limit in its energy pattern.

Also going back to the leads as each of these leads is connected properly and respectively one led to the vinegar (positive outflow) cylinder in the Joe Cell and while the other led is connected to the salt water (negative inflow) cylinder in the Joe Cell, now it is connected to a siniod coils rings that is wrapped onto the Joe Cell cylinder. Then there is a parallel connection to outlead and device from the siniod ring coils on the Joe Cell. This parallel connection can be connected to the device and car and home that you wish for the Joe Cell to power it. the Joe Cell can now be the zero point energy power source for anything and device that you desire as well, remember the torsional fields that charges the Joe Cell is from the RainMaker device, which the RainMaker uses zero point energy coils that is the siniod connected to the aluminum copper bismuth scalar coils which connect to 90 degree from the sinoid coils.

That means if the siniod coils is east to west in position so the scalar aluminum bismuth scalar coils will be north to south in position to the sinoids. This 90 degree position will create torsional fields that run the RainMaker and the Joe Cell. In turn, the Joe Cell created emf from the torsional fields that run home, car, and spaceships etc. Also this combination of RainMaker device and Joe Cell can also be used to facilitate stargate meditations and cause stargate travel.

Now don't forget the quartz crystals sphere can be added to the Joe Cell by using hot glue stick and glue guns to mount it on the Joe Cell top caps and also it conical caps the crystal will interact with human consciousness and minds when exposed to torsional scalar fields generated from the RainMaker device and Joe Cell device combinations.

Now as for the consciousness ship, again, the construction has no limit in design.

You have the quartz crystal, surrounded by the isoca-dedecahedron magnetic ferrite or neodymanium, then both quartz sphere and magnetic platonic magnetic array is surrounded by both iron and copper petal hemisphere. Then two sinoid and 4 scalar coils is connected to each other in 90 degree position from each other either in series or parallel to create torsional fields generation which allow consciousness interaction between its crew and ship consciousness. There are also two obelisks which either a siniod coils or scalar coils can wrap itself on to these two obelisks one in the north pole of the sphere and the other obelisk is at the south pole of the sphere as well as the ships hull is made of aluminum/bismuth metal/electrical tape (dielectric material) /aluminum metal. Or you can have this combination of metal hull, aluminum metal/bismuth metal/ electrical tape (dielectric material) /copper metal in this two sandwich metal and dielectric combinations is set up like the torsional capacitors. No limit whatsoever.

Now going back the Joe Cell and the RainMaker combinations, remember these two technologies are one and the same they are both torsional field consciousness interact technology except that the RainMaker device is more user friendly then the Joe Cell system but both systems are one and the same. Now once you combine these two systems that are the RainMaker device and Joe Cell device you will notice that the RainMaker device will be the control system and consciousness interface for the Joe Cell device. The Joe Cell is very unpredictable because it is very sensitive to the curry earth layline and the cosmic hatman line so that is why it behaves very unpredictably because of the hatman cosmic layline and the curry earth lay line influences it. But with the RainMaker device added to the Joe Cell device then the new combinations of Joe Cell and RainMaker device become very stable and predictable in operations and in consciousness inter-actions for zero point energy production to running both space ship and car, boat, etc. No limit whatsoever.

Now as for how to use the ship you must apply the stargate meditations method from the stargate ascension books is at Amazon.com, Borders.com and BarnesandNoble.com.

As well the light body method described above can be used by the facilitators inside the consciousness drive ship or near the

RainMaker device or near Joe Cell device. The facilitator can facilitate both themselves and traveler as well the consciousness drive ship and the RainMaker device and Joe Cell device into another density of reality or another phased dimension physically. No limits whatsoever.

Now A Word From Dave To Kosol

Kosol,

I took a look at the ion RainMaker machine. I see two things present that may give it scalar or density type abilities. Things we have already done at different points of experiment.

1 - The otagon [platonic] form. There are 8 mirrors setting across from one another. IR light will be bouncing in a platonic form across the main beam moving outwards. Since there is only one frequency of light present, these crossing beams may tend to spin the big mirrors light photons along their E field vector. If this machine is creating electrons from light photons this would be a very important discovery. A huge Ozone generator.

2 - The unit may be operating as a torsion generator using light. Just like Martins fiber optic cables, where we see a good strong field can result from light alone being directed into "opposition" or opposing flows of energy. The side mirrors deflecting light at one another. So long beams of light crossing in opposite directions can create scalar torsion fields.

3 - Using lazars could be expected to make the field much stronger, however in keeping it spread out the effect will shoot into the atmosphere and may be creating Ozone, similar to what the sun normally does to the upper atmosphere. This is a natural effect in the high altitude layers of the atmosphere. It could be expected that producing Ozone in the upper layers may be beneficial. It is questionable whether generating it in the lower layers is. During lightening storms Ozone can become very strong, so this is probably not a bad thing to do if it promotes rain.

Very little information is available on the theory, or if there is a theory, and this is not helpful for such challenges you are proposing for a RainMaking race to be effectively started. It is however likely from the vedios that if both these techniques were combined one may reach a level where clouds can be pulled in from the ocean, using our torsion field RM units and tube devices, reversing them for equal time periods, and then shooting the clouds that move in with Ozone transmissions to make them drop their loads.

I would like to see one of these in operation and run a hand around it, to see if torsion fields are present.

Sincerely,
Dave L

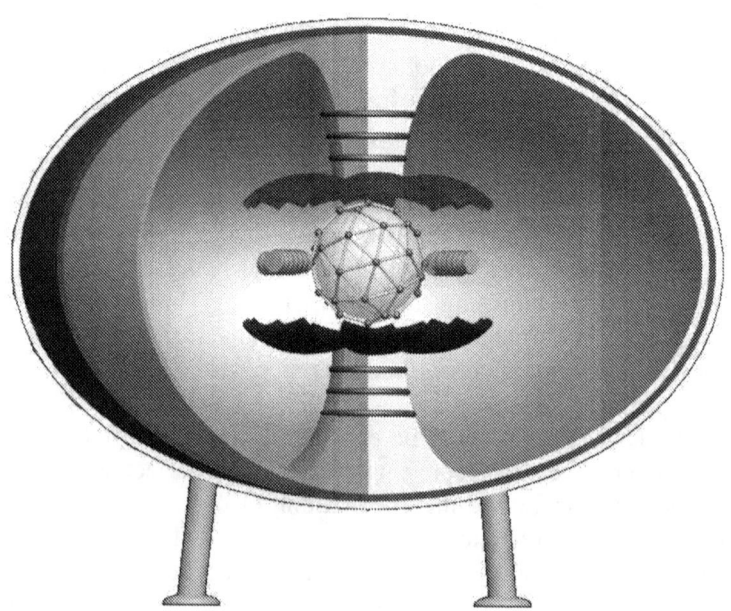

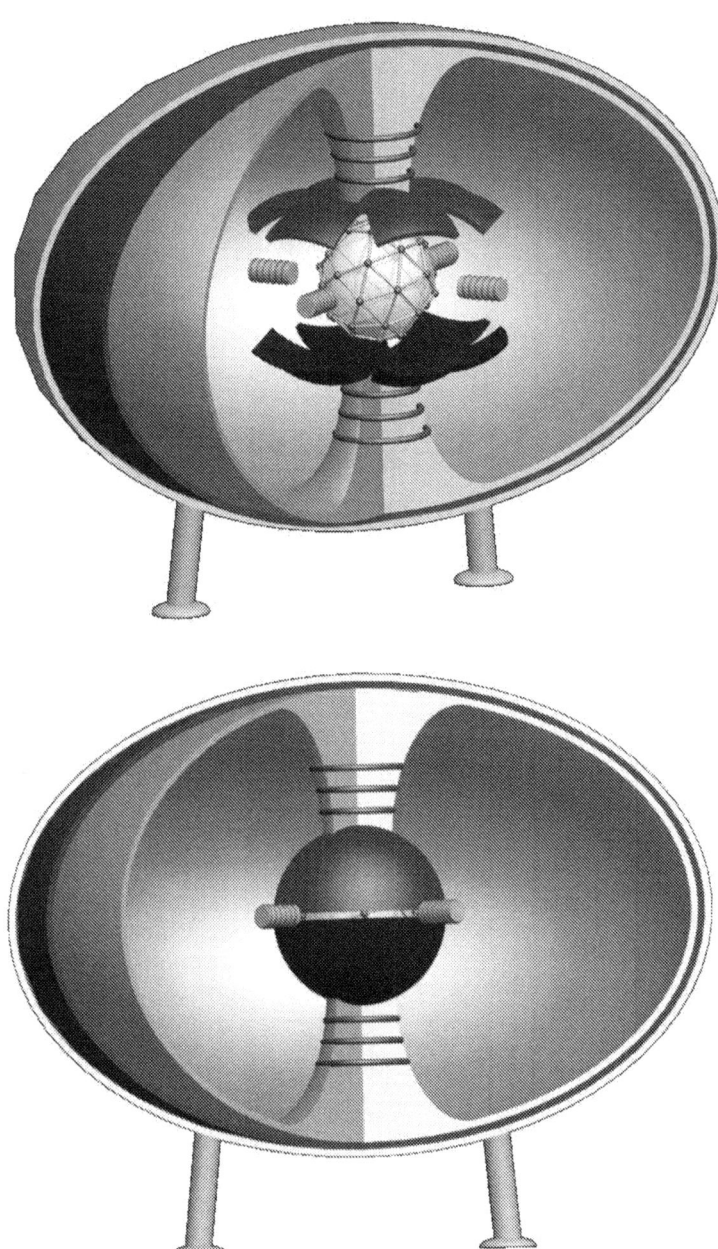

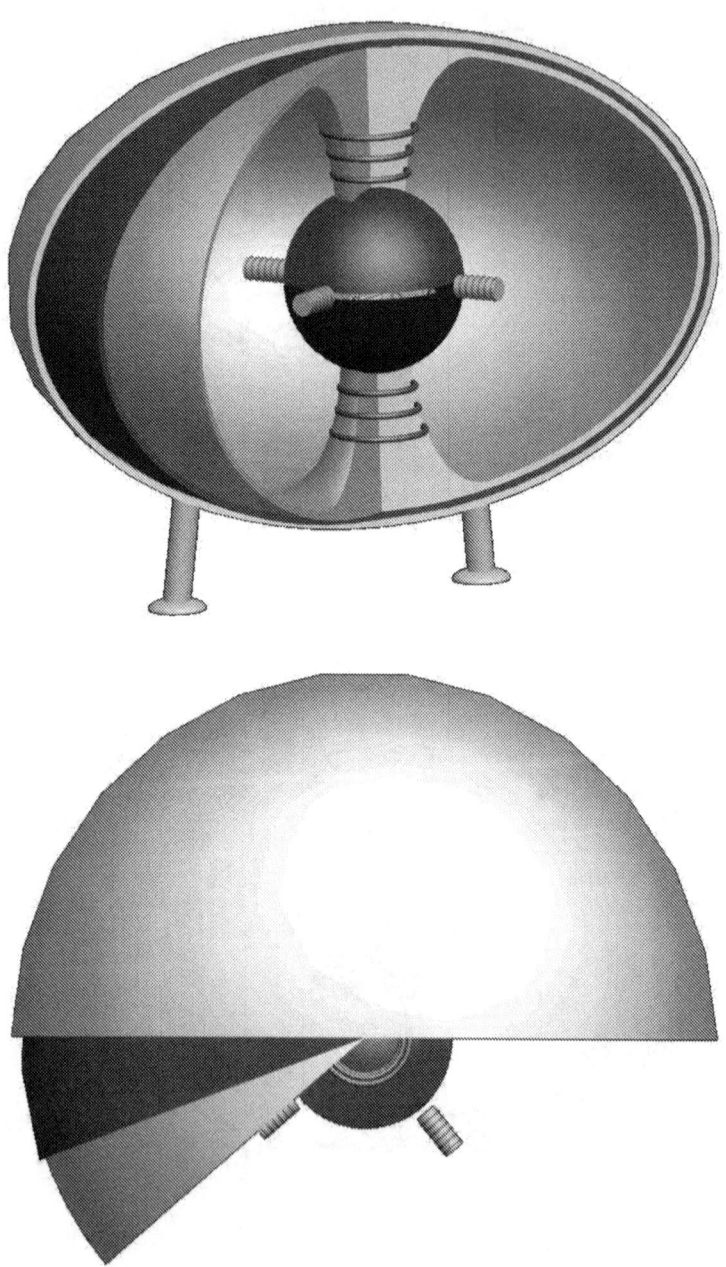

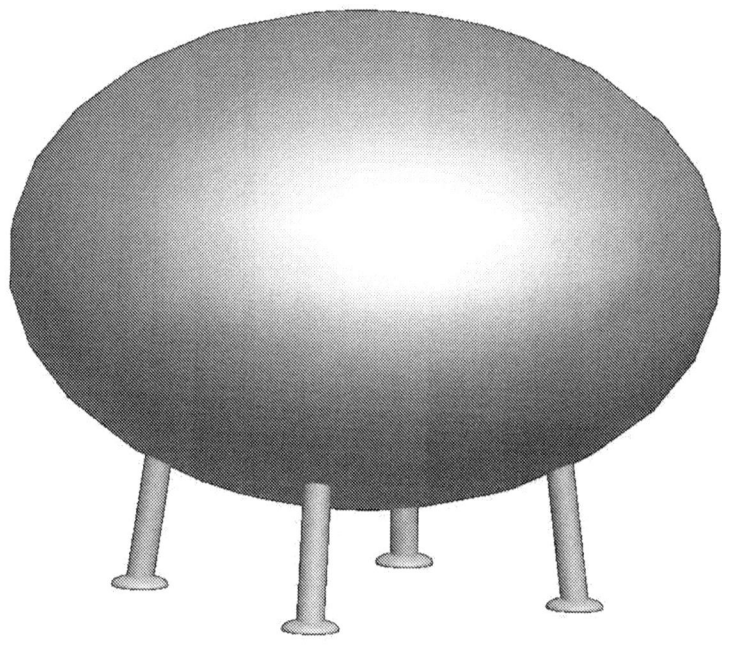

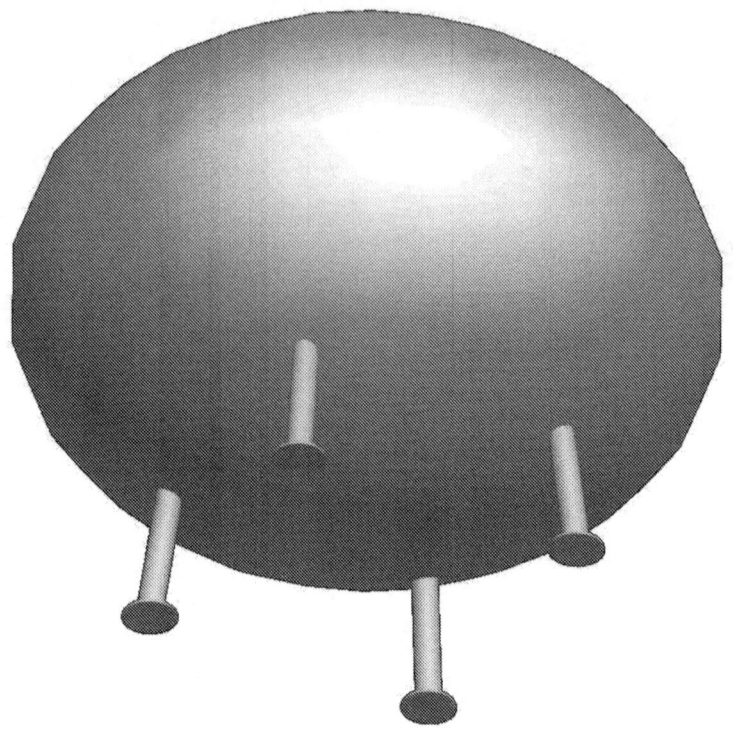

Guardian Message

Kosol Ouch Tue, 21 Nov 2006

The guardians have messages for Dave, Vince, Deltas, john, Francios, Debras, Dell, Martin, and any other builder. Here is what they said:

Kosol: Guardians go ahead and say what you need to say and I will post it to the group.

Guardians Engineers: Dear one, we of the guardians have been watching you all from our mother ships, our ship are here in mass.

We have asked Kosol Ouch the universal translator, whom we engineered 33 year ago, as a weapon of mass destruction and of creativity to send to your worlds.

He was to serve as universal translator and as a conduit between our ascended species and yours.

We have and continue to protect him and guided him to the present level. Through him we can assist you in bringing all of you to the level of nature consciousness technology to your awareness and creative practically.

That has now happened. Our ship has been monitoring the consciousness collective flux has jumped to a first contact requirement. That is why you dear one was given the ship design to be built in a short amount of time with home made material that you can obtain in your local material store and online store in your physical collective networks of consciousness call the world wide internet.

Dear one, you do not need any more mechanical technology. We now have given you all nature consciousness technology principal and practicality.

115

We of the Guardians are your creators and have always been watching you through the fabric of space and time. We are the ones who seeded you on this world and protected you from the galactic reptilians alliance call the Anachara Federations who came from another parallel and alternative universe 3 billion of your years ago.

The reptilians have many species as their slave, from the grey to the inhabitants of the known galaxy.

We of the Guardians are ascended beings and species. We seeded you on this world 900,000 years ago. Others of our kind seeded the first human colony close to 2 million year ago, before the reptilian Anachara alliance destroy it, then we reseeded your species 900,000 years ago. A place you call Lemuria continent between America and China in the Pacific Ocean.

You dear one are our descendents, we call you the 8 evolution of our legacy.

You dear one belong to a cosmic family called the Universal Federations of Light.

Your other brothers and sisters live in the inner earth of your planets. Called the Shambala Kingdom or Argathta networks as well your other brothers and sisters also live in the ocean city around the world. A place call Devil's Triangle, Bermuda, and many other oceanic grid points.

There is dear one many inner networks of both inner earth city and oceanic city network that interlinks with each other. That is why dear one you see our ships go in and out of the oceans, you call that USO, and when you fly you call that UFO.

Now dear one all of this ship are technology that is 27 million year old technology. All of the craft is nothing more then scalar field and diamagnetic field crafts. As you can see there is solid state version and also mechanical version.

They all do the same things dear one, which is to produce diamagnetic field and scalar wave field that allow the operators and crew to interact with the ship.

Now dear one we has given you our technology the basic principal and practicality that we of the Guardians civilizations use.

The ship we give you is a weaponry ship. It will teleport you to any known and unknown locations of your preference. One you link with the ship, he craft will allow your consciousness to see the fabric of creations from here you can go anywhere by just thinking about that location and desiring to be there.

Dear one there is no such things as impossible when you are using nature consciousness technology. Our technology is simple and can be built with material of your worlds. It doesn't need any nuclear material, just need your basic aluminum metal, bismuth metal, crystal quartz sphere, aluminum bismuth scalar coils, frequency generators, and finally "you". You are the most important part of the ship for without you the ship will not operate or work. So you see dear heart, the ship is a extension of you and your consciousness/minds.

We know dear heart you have many question, but you don't need to ask, just build what we have asked of you, then the answer is given. The reason we have not given you dimensions, is because we wish for you to learn and also to be creative in your dimension design for your craft.

Some of you want to build big ones to carry 20 or more people and some of you want to build little ones for two persons only, so you see dear one we know what is in your mind and heart. That is the reason no dimensions are given, we leave that up to your creative mind to play with.

Dear one there is nothing more to said except that we will wait for you to come to the ship. We have prepared an earth like environment on the ship for you and also a community that if you wish you can live with us. Your planet dear one is going to change into a crystal planet and you dear one will live here with us in our ship and on the fifth and six density and beyond.

So dear one please build the ship fast, you can alter the design and shape of the ship to fit your likes. As Kosol tell you, there is no limit whatsoever.

We just give you the ideal shape design but you can build your own shape of your desire, no limit whatsoever as long the natured consciousness principal is followed.

Remember this craft is one with you and will do what ever your heart desires. Remember dear one it is not only a transportation craft but also a weapon of mass destruction and creativity. Be careful how you use this craft once build and this craft responds to love, loves run it and love is the technology and energy behind it and you and God are love.

We wait for you.

We love all of you dear heart.

Kosol: Thanks Guardians. I now forward the messages to them.

RainMaker Discussion

Kosol, Peter,

Don't like making comparisons but we have distinctly two entirely different means of creating rain here with the AIR device and the RainMaker 1.

They both fit into what we require to make it rain but they are of two Worlds and each has its respective and responsible place in society.

I have about 8 proven methods of making it rain and will one day put this all together as a PDF for others to use.

The first, we are discussing here, the AIR (Atmospheric Ionization Research) device of Peter Stevens and Jack Toyer we can equate with known and current technology and should be snapped up by Governments because it is not an Alternative style of Rainmaking. It fits in neatly with current Science teachings.

Peter needs to put a huge $$$ cost on this device and this will ensure it is utilized by Governments.

A new Government Department needs to be set up that will cost Millions of $$$s to implement this technology.

I have witnessed the AIR device in operation here from my daily Weather downloads and am truly impressed.

Out of season Monsoonal style rains to Northern Australia (WA, NT and Queensland) which set records in most areas.

Most impressive!

I am fully supportive of Peter Stephens AIR device and is a known current day Science solution to Drought and should be immediately utilized throughout the World.

This is on par and possibly better than both the Wilhelm Reich or the T J Constable devices utilized in rainmaking.

In fact I have an AIR style device here, which I developed several years ago but was in a pipe arrangement and probably needed a dish/cone style aperture to spread out the contents instead of a focal point.

This device was never fully tested as I was using a rotating TJ Constable device which was meeting my current need and perceived successful.

However, again Peter has a difficult job ahead of him to influence current Science of his or Jack Toyer's device.

Current Science cannot explain why 'Clouds do not fall' nor can they explain the true polarity physics of the Atmosphere and why it does or doesn't rain. While you continue to deny the presence of an Aether you will continue not to understand the true relationships of the mechanisms that support rainfall.

I believe I have studied this subject fully and understand the above mechanisms and would like to put that also in a PDF for all to see.

I plead with interested people to read Victor Schauberger's 'Living Energies' by Callum Coates to be able to grasp some of the benefits of Vortex or Implosion technology as related to our Living Earth and Atmosphere.

The second, 'RainMaker 1' (RM1) is different - it can clearly be labeled as Alternative, Metaphysical and using technology which is not accepted by Science and Governments. Anything that reeks of Witchcraft, The Aether, Implosion etc and hints at something other than the current Science will not be tolerated.

There are no winners here! Both are winners and both have their place in this World.

One accepted by society (hopefully?), the other definitely not.

That is my Diplomatic answer to your question.

However, what makes the RainMaker different?
This is difficult to explain.

Firstly we cannot really see what we are doing or creating and we need to understand why first.

We do not have any measuring 'Tools' but only our perceptions and senses.

Most of what we do is a perception that we believe we have created 'something' but what?

Our perceptions are that Planet Earth was in need of a revitalizing energy of some sort to bring it back to 'normal'.

As individuals we need to develop our sensitivities, our Common Sense, our 'connection' with and to Nature in an attempt to understand what it is we are achieving.

I believe this is the very first step, like a child learning to walk for the very first time and this is where we are at in terms of learning and understanding our new found technology in the RainMakers.

First requirement is to understand fully, your subject you are working with.

My case was easy, Australian Weather - the driest Continent on Planet Earth.

To date in Australia, we believe we have partly returned to 'normal' after a 17 year Drought and perceive the RM1's have been responsible for this turnaround. There are 3 in use here.

The current means by which the Government decides whether we are in a 'LaNina' (Wet) or a 'ElNino' (Dry) situation, the SOI (Southern Oscillation Index - air temperature comparison Darwin/Tahiti) does not indicate improved rainfall to any degree whatsoever.

Current SOI is at minus 8, very much 'ElNino' still!

But I say 'we are partly back to normal! Southern half of the country at least. Central strip which includes SE Queensland is the major area in need of improvement.

People need to understand that to get your dams and rivers filled you don't get this with day to day sunshine and storms and winds are part of what is required to make it rain to fill those dams and rivers.

This means there IS going to be floods, wind and rain damage, hail, lightning etc and possible death because somebody was out of tune or touch with Nature or just plain stupid.

'RM1' can be utilized in two ways:

Firstly as a RainMaker: Used in either a North or South Out mode to promote Fine Weather or Rain. (Reverse in Southern Hemisphere).

Secondly the 'RM1' is a Vortex creation device that can be manipulated as a Radionics machine by using an individual's 'Conscience' energy and 'Intent' to achieve a certain or planned result.

Wherever a Vortex field is present, we can impose upon that field our 'Conscious Intent'.

The 'Tube' device is still a RainMaker, directional but not reversible.

Two Tubes have been used to couple energy from an Indigenous Spiritual Centre here in Australia (Uluru - Ayer's Rock and Kata Tjuta - The Olgas) to an area of the Riverina, in southern New South Wales, badly in need of rain.

A second directional thrust in January 2007, with one Tube was used to bring record rains to an area ('New Crown Station') of Central Australia bordering on The Simpson Desert.

This was in response to a plea from the owners via our ABC (Australian Broadcasting Commission) but was made in total anonymity.

(This was different to an AIR event which is dragging in moisture from the Indian Ocean in a Trough but an Upper Atmosphere event at the 500hPa level as a circulating Trough or Low Pressure system.

Both technologies clearly have their own signature.)

Both of these Tube events were perceived successes using 'Conscious Intent'.

The latter is now subject to a revisit to confirm the first effort was not just pure coincidence.

Both of these will be formally prepared into a PDF for public use when completed.

In Summary:

Two proven RainMakers but belonging to two very different Worlds that are truly Worlds apart.

Trusting I have made it clear that there is no comparison here in the divided World we live in at this time in History.

Appreciation to Kosol (Kosol Core Tech) for first describing this device and then to David Lowrance (CSSP) for implementing Kosol's plans and noting its ability to create rain and making it so easy for me to replicate and adapt into the Tubes.

There are untold further benefits available from the RainMaker and we are currently researching those benefits.

This is what makes the RainMaker different and unique and a new understanding is required from all to accept this new technology.

This document has been in front of me all day - it is a difficult question you have asked Kosol but hope I have answered fully to the best of my ability.

Thanks.

Smokey

Hello all here is the ultimate RainMaker device, it has zpe coils and a frequency optional connectors so it has it own power to run from as well it can also use the frequency generators for experimentation.

As you can see, I have two crystal spheres, quartz, one big and one small. As well there is a hot glue chassis that is cooled off all ready. What I did was put the two sphere quartz in the freezer to get ice cold then I bring hot glut to it and pour it to the side with the glue guns. Then the glue will stick temporarily to the quartz sphere then a minute later the glue cooled off and hardened into the form of the sphere quartz crystal but will no longer be attached to the quartz sphere so you can removed because of the coldness from the quartz has condensed the warm air into water and moisture.

Also the zpe coils consist of a siniod 24 gauge copper wire coils on a iron bolts, then a siniod coils on a aluminum foil pipe that I made, and then a sinoid coils on a copper pipe connecting into series to each other. Then is connected to the zpe coils of the scalar aluminum bismuth copper coils which consist of 24 gauge copper wire coils 4 way scalar wrapping on an aluminum tube and inside this aluminum tube is the bismuth metal which I melted and poured inside.

Once this is done the scalar aluminum bismuth scalar coils is inserted into a copper pipe, which this copper pipe is also will be wrapped with a 24 gauge copper wire in a 4 way scalar wrapping formation. Then these two scalar wrapping coils both on the copper pipe and aluminum bismuth tube will be connected in parallel connections. Then after that it will connect with the series connected siniod coils that is mentioned and described above.

Now you got your zpe coils and it can run itself. It will run your RainMaker device with torsion. As well you can connect the frequency generators to it also it will give it more power.

And now you also have the option to run the regular copper aluminum bismuth scalar coil which is located inside the iron ferrite so there is a total of 3 coils in all, one siniod, and two scalar coils. One is inside the iron ferrite and one is attached to the side of the crystal which is 180 degree from the siniod coils that is on the other side of the crystals.

Then the ferrite or neodymanium magnetic is used to create the inflow or out flow. Inflow north turn inwards and out flow north turn outwards. The torsional fields is very strong from this coils. Remember the siniod coils are connected in series to each other and then the siniod coils are connect in parallel to the aluminum copper bismuth scalars coils.

Just remember you can now use rock, wood as well as anything you want to let it sit on the RainMaker device beside crystal, so you can experiment.

The device can now run on cosmic and earth energy fields thanks to the 44.4 feet wire that is used to create the siniod coils and connecting that to the scalar coils.

So no limit whatsoever. Don't forget these coils must be ninety degrees from each other in placement, that means they must form a cross. If the scalar coils are laying from east to west (horizontal position), so therefore the siniod coils must be in a position placement from north to south (vertical placement position in other words) or if the scalar coils is vertical placement position then the siniod coils must be in the horizontal placement position. It must form a cross formation of these two different coils so you can have a 90 degree placement position from each other to create and catch the torsion field which can run the RainMaker device, so there is no limit whatsoever with this consciousness technology device.

Regards,
Kosol Ouch and the Guardians

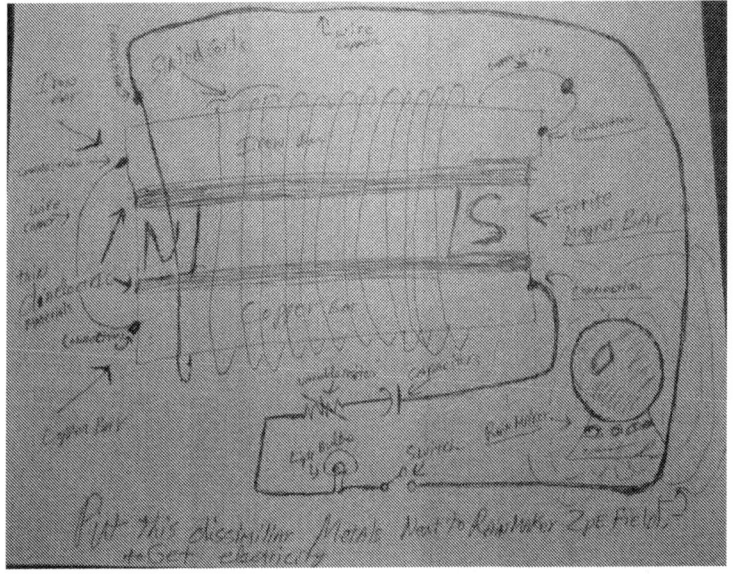

127

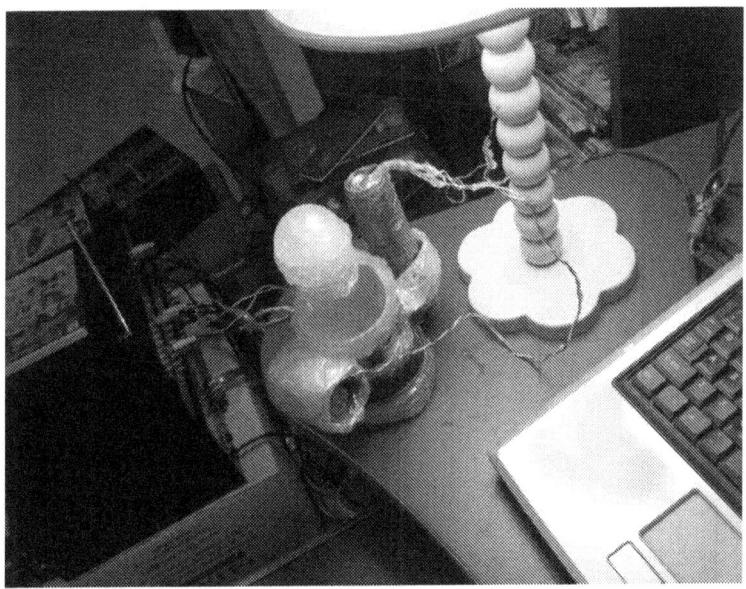

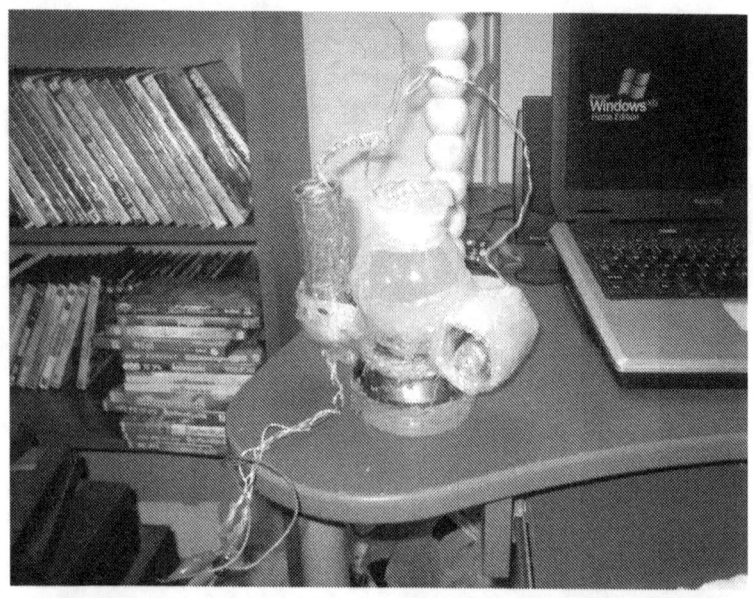

Group Discussion

LOL!

Tanjmaster, this is one of the wonderful things about being first to build a device. You get to engineer all the problems as you go, and everyone does it different! How do we get these materials and fields to interact? What to mount it on is very important, I would be inclined to use plexiglass cut structures myself on this stargate one. Time consuming for sure, but what an art form! LOL!

I see Javier's device is also using some stacked coils like this, and his base is not made from TV parts but some authentic magnetite minerals of some form.

The pdf "Flavio Thomas Pyramid" is now fully understandable. This has answered some of my initial questions. Thanks so much to all that translated and presented this information.

I am starting to see a common theme to the Joe Cells, the Carr Utron, and getting EM from torsion fields in Flavios pyramid device. Opposing diamagnetic torsional spin fields being pulsed by a magnetic field. In the case of Flavios unit no pulsing to produce DC to run a small fan motor, just a magnet on one side to unbalance the device.

Rather than using dissimilar metals he would seem to be placing opposing torsion fields on a copper capacitor, a sort of torsion capacitor. The two coils, one CW one CCW to alter the pyramids torsion, one up and one down to plates on the cap, and each one is a different number of turns [unbalanced]. Now whether the magnet is effecting the upper coil, or the capacitor is unknown, but the similarities to what Kosol is seeing with the dissimilar metals is striking, as the Aluminum and copper will offer an imbalance as well. The iron brings the energy back into the electron shell. [Remember the Utron, opposing torsion fields joined with a magnetic pulse.]

131

I can't help but feel we must be getting close on this, only more experiments of course will eventually get us all there together. The one thing I would suggest is that one does not need to build a full sized pyramid to generate a torsion field anymore, as a density sphere will do this and much stronger. Unless you want to experience setting inside it of course! LOL! The torsion can be routed anywhere with a single wire, even to one plate of the torsion capacitor, the other two plates having a torsion differential based on the two coils setting below it. Seems like some good experiments are coming to mind now on this.

Kosol,

Congratulations on your new creation, I'm glad that it is working without the funct generator, and it is a stronger unit. The coils look interesting and would seem to have all the right stuff! It will be interesting to see what the following months hold in the alternate energy field.

Conversation With Kosol 7/26/2007

Dave L and Kosol [reflections and where we are presently at].
[I was instructed to take notes.]

On Torsion Fields:

Torsion fields are attracted to platonic form.
Torsion hits sphere - pure consciousness
Torsion hits pyramid - electrons
Other platonic forms - other functions - density travel etc...
[Torsion fields have layered shells, at higher levels become sheering forces]
Suggest winding CW on iron - CCW on Aluminum [as both spin opposite directions this may have a similar effect on both materials]

Water From Crystal:

Last winter I had an experience with a crystal sphere where it seemed to be seeping water, I would dry it off, but the next touch felt like a new layer had formed. The sensation was amazing, in that there was no way to get the sphere dry, as though the water was coming from it.

Kosol related an instruction from the Guardians - place the sphere in the freezer for about 15 minutes and get it down to 50 to 60 degrees. Wrap the sphere with aluminum, poke a hole in the bottom and collect the water in a cup after you take it out of the freezer. The water is charged with an energy field as well.

During my experience here in Alaska after shutting the window and warming the room, I viewed this as a unique but annoying side effect of cooling the sphere. An experiment could probably be set up with this to see how much water can be released using this method. While one immediately assumes that the water is being pulled from the air by condensation, I must admit it did not seem this was where the water was originating. Also if this could be done at very low humidity say in a dessert area, it may provide a surprise or two. Kosol related from the Guardians communications that the spheres can produce water if cooled either by normal temperature or by strong enough inflow, and this could be a valuable commodity where no water was available. After witnessing this myself, I feel it may be worth investigating further.

Conscious Tech:

The discovery of torsion fields, and scalar coils, being somehow linked to consciousness itself, now some years back, has led us down a unique path here on the Kosol sites. The outlandish claims that at some point the Spiritual and the Scientific would come together, by this little Cambodian man on an antigravity site, I now caringly refer to as "little brother", a message not well received by many at the time but one that lit me up with a new vision almost instantly. Well now we have "RainMakers" and "dodeccahedron healing orbs," ZPE torsion coils, and conscious interactions with "scalar coil machines", it is only beginning, the baby steps.

Outlooks:

The clue is that if we can get torsion fields to interact just right, then electric fields will come out, and the energy will be coming from the force that causes matter to spin. The force of light itself, and here is why there is a conscious connection. This is very hard to communicate if one has not experienced it firsthand however. The message I get is that the Light simply wants to be recognized as the "source". [Using torsion fields to generate EM is getting the back EMF without using the forwards EMF to generate it, but directly from the intermediate torsion interaction of the nucleus of copper atoms.]

Catching ghosts? After my graveyard meditations many years back, I think I would set the RainMaker on outflow send the spirits "Light" for healing whatever emotional turmoil may be holding them here. Most of what I got from "reading ghosts" in the grave yard were deep emotions and sorrows. Many are already in the inflow state on the emotional layer. However if one wants to suck them in and take them home then please be my guest and set up for inflow! LOL! Use a camera on infrared night mode and get some shots, you may be amazed. The RainMaker will almost disappear on inflow mode just like turning the third eye center inwards will cause someone to seem dark as if a shadow came over them. Reversing this however will fill an area with "Light" and there is little emotional deception possible when standing in the Light of a conscious device.

Reflections:

The RainMaker device is producing and amplifying what we all carry inside us, a powerful chakra system on 7 levels of consciousness. We can slide between these levels naturally as humans and usually don't explore them all until late in life, maybe in our 60's if we continue to grow all our lives, and do not get stuck at one level, affixed to the power center at the solar plexus most often where the major part of mankind is setting.

I sincerely hope that in my lifetime we will reach sanity at the low heart center as a collective. [Kosol relates an incident where a conscious device caused division in people, that become almost explosive. The division seemed to be between those who could sense the energy and those who could not.] Each side was atament, but those who could not became emotional and blocked the energy flow which seemed to escalate there emotional state.

I know how this happens because I have seen it occur with very spiritual people present in times past. The presence of a strong Light energy brings up all that needs to be healed, and some are not ready to face this in themselves. Rather then recognize the turbulence is within themselves they project the cause of their turbulence on others that are present, even claiming they are the source of evil. Basic therapy teaches to recognize who is having the turbulent emotions, and teaches to own our negative emotions and not project them on others. I believe this should be taught to all teenagers as the emotions are developed in life and coping skills given, but then so many in our society are still locked into ideals of God being a ruler with iron laws rather then understanding ourselves with self love as the adepts have all taught.

Anyway, I cannot imagine being in a room with 50 people and firing up one of these conscious devices as a demonstration. Half may feel good light and half may accuse you of being demonized! LOL! Makes me wonder. But the units do accelerate the meditational responses, if one is desiring it. Much like the Sedona vortexes, and deep state can be achieved in record time for those seeking one.

Manifesting:

Kosol relates the formula, "Move into the higher vibrational state, create the form, bring it back down to our density by lowering its

135

vibration, it appears in the physical." I must admit it still seems like magic to me, and I personally cannot conceive getting to this step in this lifetime. I am barely beginning to see how it may be possible to manipulate torsion field to produce EM! Wine... Water... water to wine! Gotta wonder, but if Kosol is seeing this stuff I take note and listen. The most I got was a syrupy feeling on my fingers that felt astral and gooey.

Conclusion:

Each time Kosol prints up a book, I sit there and think "but it's not finished yet!" then I get one and read through it and remember where we have come. I don't know if others are finding joy in this process, but I certainly wouldn't change a thing.

Kosol,

Thanks for the continued vision, the written record, and thanks for the chat tonight,

Dave L

How to create living water by introducing cold temperature to the crystal quartz sphere of the RainMaker device.

The concept is easy, all you have to do is put the crystal quartz sphere into the freezer of the refrigerator for about 15 to 20 minute or until it gets very cold, then bring it back and set it inside an aluminum foils or copper foils spherical container that is bigger than it and thick enough to hold its weight. This copper or aluminum foil must have many little hole punched into it by a small nail. Then after that, water of living aetheric high charged quality will pour out of the crystal that is inside this aluminum or copper foils container. The water of high aetheric charged quality will drip and flow out of the container that crystal quartz sphere is in, to fill your glass so you can drink it.

As long as the temperature is cold between 60 to 40 degree on the surface of the quartz crystal sphere, you will get water pouring out from the crystal quartz sphere to fill your glass. Now the aluminum container can be made of aluminum foils or other metal such as copper container but they all must have little holes in them so air can come through so the cold crystal can convert it into prana or chi or aetheric life force water for usage for people to drink to heal and strengthen both spirit, mind, and body with this highly charged aetheric live water that comes from cold quartz crystal sphere.

Now as for the torsion fields generators from RainMaker device, the torsion are attracted to plantonic form. As well it will amplified that for example, torsion field generated from the RainMaker device will be attracted to pyramid structure to produce electricity etc. As well it will attract itself to human beings and isocahedron, dedecahedron, square, tetrahedron, octahedron, isoca-dedecahedron plantonic form as well as tree, rock, animal, fish, water, etc. This torsion chi field will encompass that form and will interact and empower that particulars plantonic form to create electricity, time travel, antigravity as well as hyper drive. No limit whatsoever.

1. Now as for the RainMaker device, to run it without a frequency generator, you must shorten the bismuth aluminum copper coils. Shortening meaning you can put the wire of one end of the lead and connect to the other wire of the end leads of the other scalar bismuth aluminum copper coils with in that same single coils units.

2. Another way is connecting the bismuth scalar coils in parallels to bismuth scalar coils and then these two coils can connect to a siniod coils in series. Now as for multiple siniod coils you can connect in series to each other then connect them to the bismuth scalar coils in series.

3. As for the density sphere coils you connect them in series with each other, then you can connect them to a siniod coil in series.

4. As for the scalar bismuth coils you can connect them in series with the siniod coils after that you can shorten the density

sphere coils and use one wire from the shortage density sphere coils to connect to an open side connection of the bismuth scalar coils and siniod coils.

I know this sound crazy from the circuitry of electronic, but we are not dealing here with science only we are dealing with spirituality that is torsion field technology. Remember the 33.5 feet of 24 gauge copper wire creates good coils for siniod as well the 44.5 feet of copper 24 gauge of wire create good siniod coils for the bismuth scalar copper coils to be connected to.

5. Now you can do experiment to see which combinations is good. You can connect the shortened connections of the density sphere to a shortened bismuth scalar coils and then connect that to one end led of the siniod coils while the other end lead of the siniod coils is connected to ground. This will give strong torsion fields, so turn off you frequency generators and oscilloscope, voltage meter, amp meter, etc. for the only things that you will need is your palm, mind, body and spirit to detect torsion fields and see how this torsion field affects weather, gives electricity, etc., as well create antigravity, etc. No limit whatsoever.

6. Remember all of you are torsional field technicians and torsional fields engineers. The torsional science which the consciousness has no limit. Its circuitry defied the circuitry of electronic totally there is where science meet spirituality which consciousness torsional aetheric or chi fields harnessed into technology. Very importantly that the crystal quartz sphere has the ability to amplify the torsional fields because crystal quartz sphere or any type of quartz in any plantonic form or in any other form amplified the torsional fields created by the bismuth copper aluminum scalar coils, density sphere coils, etc., so no limits, because crystal quartz in any form has dual torsional spin from proton to electrons to photon etc., as well as consciousness spin on both side counter clockwise and clockwise rotational spin. No limit whatsoever.

So remember to experiment and shorten them bismuth scalar coils to get torsion fields or you can connect them bismuth copper scalar coils to the siniod coils to get the torsional fields etc.

Hehehe. No limits whatsoever and welcome to consciousness technology model principal practical applications.

7. Now as for materializations all you have to do is raise your consciousness vibrations to the celestial thought frequency and then you can create any things you want by thought then after that lower that particular thought form frequency into the physical density that you just created then it will materialize right in front of your eyes. So that mean you can create anything you want. The RainMaker device, meditations on the center of the sun or even use stargate meditation help to raise your frequency vibrations. Meditations on the center of the brain sun is nothing more than just visualizations of sun with all its brilliant radiance colors the size of a golf ball in the center of your head or brain. Then you just breath in through your nose saying mentally in, then breath out through your nose saying mentally outs at the same time you visualize the center of the head suns. If you visualize a line at the top of your ear and draw that line touching the top of your other ear, then you draw another line from the center of your forehead where the 6th charkra or third eye is located, and draw that line from the third eye to the back of your head in a straight line where ever the line from top of the ear crosses or intercept the line from the third eye and the back of the heads. Then that is where the center of the brain sun should be located that where you visualize the center of the brain sun. Its physical location is the hyperthymus gland which contains the pineal gland and pituitary gland, which is the gland that produces hormones for everything in the body. No limits whatsoever.

Regards,
Kosol Ouch and the Guardians force

The Earth is going to be a monopole because of the photonic belts so all electromagnetic fields technology will cease to function. The only technology that will function is diamagnetic torsional consciousness fields technology.

Here why you all should pursue RainMaker technology. The Guardians said that the earth will become mono pole and only torsion diamagnetic consciousness technology will work. All emf technology will cease to function thanks to the photonic belts affect and the earth becoming mono pole as it reaches full consciousness.

Torsion fields are attracted to platonic form when there is pyramid form present or even tetrahedron platonic form the torsional fields will create electricity. When there is a square platonic form present the torsional fields will flow into that form to create a fourth dimensional field portal if a person or object step into this square form he or she will step into the fourth dimensional reality or density. When there is the platonic forms called isocahedron platonic form there will be hyperdrive teleportation effect when torsional fields flow into it when any one who step into this isocahedron form. Where there is the dedecahedron platonic form the torsional fields will flow into it and will create dimensional hyper communications and transportational effect on any one or object who steps inside of this plantonic form. There is isoca-dedecahedron platonic form, when torsional fields flow into this plantonic forms you get instance ascensional physically and spiritually. Now if there is sphere platonic form and torsional fields flow into it and any one or object step into it, he or she will experience the mind and consciousness and body of God which is Lord's vishua.

Now all of you must pursue this consciousness torsional fields platonic RainMaker technology for the earth will become mono pole and electromagnetic fields technology will not work anymore, emf only works if there is a di pole but once the earth become mono pole because of the photonic belts and is increasing the earth frequency and turning the earth into a full consciousness platonic consciousness planets so therefore it will become torsional fields and diamagnetic fields so only diamagnetic field and torsional consciousness field technology will work and electromagnetic fields technology will cease to function.

So take note everyone.

Regards,
Kosol Ouch and the Guardians

Technical Overview RainMaker 1 Device

Rainmaker 1 - Observed Process Of Operation - Physical Plane Perception

For those who are not familiar with the basics of magnetism as it operates in the materials of the device from the physical science perception, I have prepared this overview, so one may get a feel for the forces present and how to effect them based on the physical adjustment of the device. There are several flow paths of energy at the physical level including the magnetic field, the isotope line, the diamagnetic field, as well as nuclear torsion or tempic field. Conscious connection is through the diamagnetic field where it may become dominant in either material.

We observe the conscious interactions first:

North poles out creates two states:

Bismuth = outflow [From device to operator]
Iron = inflow [From operator into device]

South Poles out creates two states:

Bismuth = inflow [From operator to device]
Iron = outflow [from device to operator]

This is because the Proton and Electron spin opposite directions in the same magnetic field alignment.

Interactions:

While the Rain effects may be possible for a dumb device, an understanding of the basics is far better. It allows the operator to make changes and know why.

It is good to gain a first hand feel for the different ways of consciously connecting with the device and a realization that we are all very unique. As a warning it is not a good idea for a new experimenter to set up two or more devices that operate in conflict as they interact with you. An example might be two systems with large iron layer mobius coils, where magnets are reversed on each one setting in the same room. This may manifest a powerful conflict between the mental and emotional state in people in the immediate area and can lead to sickness over time.

Operating multiple devices in one local area takes skill, understanding, and patience. In the above case we are routing Astral energy on the return loop to Source into the physical while at the same time projecting a conscious intention into the mental plane. As it manifests it will hit the outflow loop, and a high energy conflict will be set up in the physical plane. Source flows will be powering it. Indications are that we may not be ready for this just yet as it affects the brain very strongly. If you are not sure keep your devices configured the same for magnet polarities. Or keep the devices at different localities and well separated. If the diamagnetic field, or the magnetic fields share any overlap you may end up with a state of mental confusion and a major headache. You can check the conscious flow charts at some point to gain abilities at this level. The normal reactions for a beginner should be a heightened mental clarity, and increased energy levels. So learn to monitor yourself as a first or prime intention.

Those warnings issued lets proceed.

Energy Flow In The System:

1 - Iron Ring extends magnetic field inwards as one pole interacting with Bismuth coil

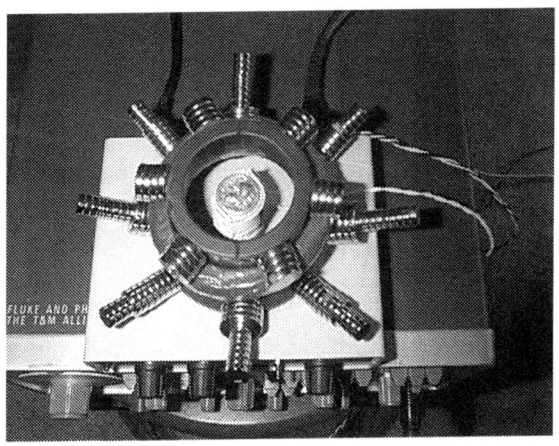

2 - Bismuth connects and begins to send nuclear torsion outwards through the magnetic field. Also the bismuths natural diamagnetic field expands slightly.

3 - Scalar coil is energized, or shorted, now on the bismuth coil and begins to deflect the energy of precession motions into the nucleus diamagnetic field. The bismuth is now interactive with our consciousness through the diamagnetic field.

4 - Scalar coil on the ferrite ring deflects energy into the diamagnetic field and reduces the magnetic field extending inwards to the bismuth coil. The iron is now interactive with our consciousness and the bismuth connection energy is reduced in level by this diversion.

5 -Quartz Crystal has a perfectly balanced dual spin quality and will generally follow the dominant field. A crystal ball however will lean towards a natural inflow state.

[If one is doing healing work I found that an outflow strong enough to move the crystal ball into outflow was very beneficial. This type of healing field is useful for me as a healing Chi source. If the process is allowed to continue for many days or weeks the CU that develops may become like a child and begin to interact. This is a natural result of bringing a large amount of conscious energy into the physical plane.]

Here we see the balances present in the RainMaker 1 from the physical perception. Altering parameters may effect the state of either material to dominate as either an inflow or an outflow.

Set Up Configuration:

To experience only the Iron, simply open the bismuth coil, keep the wires separated, and place a Tesla scalar coil or a Mobius coil or Lakhovsky coil on the Ferrite ring, and add as many magnet positions as possible long enough to reach the edges of the coil. The 12 count works good for this as they end up very close together and you may want South poles out for this [Iron outflow].

To experience the Bismuth remove all the coils from the Iron, separate the magnets to maybe a six or eight count, and use the bismuth coil with a function generator. You may want North poles out for this one [Bismuth outflow].

A device may blend both of these at various levels. If it is noticed that suddenly the magnet polarities are working

144

backwards then one has probably shifted to using the other Source flow.

Device Manipulations:

In the two layers of material the iron has one free electron and the bismuth has one free proton. These are interacting to connect across the gap through the common magnetic field present and torsion is transferred across the gap. Bismuths magnetic field is connected with its mass and Iron is not but has more power to extend its field.

As we add scalar canceling energy to the iron, less electrons are able to form the energy connection to the bismuth as they begin to cancel at the EM level with one another. The bismuths diamagnetic response to the iron is lowered, and the Source flow in the Electrons is exposed as a diamagnetic field from the Electron shell side. You can see here how the flows are diverted to the diamagnetic field. Our connection with the device through the diamagnetic field shifts to the Electron layer. This is the return conscious control loop back to Source. Manifestations implanted at this layer will be returned to Source and then manifest on the reversal.

If we remove the irons scalar coil and separate the magnets well so they do not cross precession cones in the iron, now we have a strong free iron electron magnetic field extending at full power into the bismuth and interacting more with the bismuth layer. The bismuth scalar coil now deflects this energy into the nucleus and the diamagnetic field becomes much stronger from this layer. The diamagnetic field present is now shifted into the bismuth coil and exposed more from the Proton layer. Since this layer is setting at an alternate density this effect is stronger with respect to torsion. We can operate here directly in the outflow from Source as it hits the Physical and Astral planes.

The RainMaker 1 thus gives access to both flows depending on the balance achieved by the physical setup of the device and which diamagnetic field becomes the strongest at the physical layer and whether it is inflow or outflow. Magnets thus may appear to work in reverse based on which field is dominant as the conscious link. Interaction also will vary from a verbal type communication in the physical to a more telepathic type one as we move between the flows.

I have now witnessed rain effects from either magnet positioning however it is felt that the intention is merely entering the conscious loop at a different point. As the inflow or outflow that become predominant pulls on the physical plane hardest. If inflow becomes dominant in either material over outflow in the other material rain effects may become present as the background Aether begins to be pulled over time. This effect moves directly through the tempic field.

I normally operate with no coils on the iron ferrite and work directly from the nucleus side in the Bismuth. This requires deeper states of meditation generally and a conscious "slide" into the coil. Others work from the iron side with large mobius coils out there, and this can be as simple as a note on the machine that you read verbally to state your intentions in the physical plane. Both Spiritual techniques are effective, and leads to a healthy respect for entering the Conscious loop at any point, however magnets will be reversed in each one and more scalar energy must be present in the correct material to make it dominant.

Lots of parameters here to follow, but some clues as to the different things we are seeing.

Forces And Reach:

Irons magnetic field is rooted in the Electron shell, where its mass is very low and its magnetic field has a very long reach. Iron nucleus is magnetically neutral however all its Protons are linked to the Electron shell and balanced there. It has a free Electron to interact externally from the Electron shell level. This is commonly called EM and allows iron to become magnetized coherently as one very large magnetic field that will extend many inches. Its reach will easily move into the bismuth coil at the center to align the Proton fields at that point and control their precession angles setting them into a circle with one pole outwards and one compressing inwards.

Bismuth has two qualities of importance. Its natural diamagnetic field is very weak but does extend external to the material. Thus a scalar coil close to its surface intersects this field that is already present and device motion is not necessary for this field to become interactive as it is with Copper or Aluminum.

Secondly Bismuth is magnetic at the Proton layer where its particle spin is reversed from the Electron layer. The Protons magnetic fields are all very small and barely reach the next atom, however it is coupled to its mass or weight. An isotope line may form inside it to propagate the effects that are generated only very close to its surface. The torsional effects are created at the surface of the material but move all the way through it. The bismuth therefore does not project a field outwards to the iron, it spin couples to the iron magnetic field and transfers its torsion of mass momentum through this field outwards. It also propagates this field as an isotope line into the rest of the Bismuth atoms. This is a torsion field of tempic effect and what we sense with our hands as heat or cold or vibrations from the device. These torsion fields have been associated with pyramids as well as large spinning flywheels.

Further experiment may more clearly verify this as we progress, but it would appear that devices can be designed to enter either side of the conscious flow as it moves from or away from Source. Physical properties of dominant field are however device dependent, so machines should be monitored when they are active.

These appear to be the steering capabilities for the conscious flows in the device.

Map Of The Conscious Flows:

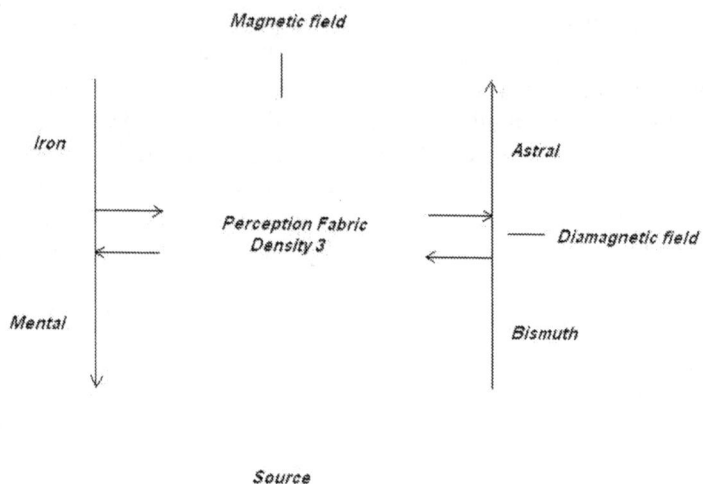

From the chart it is good to realize that inflow and outflow are terms relative to where your awareness is setting. In our normal waking state we are setting at the center of this chart. Astral and Mental planes are projecting the sphere of physical space, and we are projecting our consciousness into our physical body where we think we are setting. The flows are present in all matter, but in the Iron and Bismuth they can be hooked magnetically and manipulated.

Definitions:

Diamagnetic field: Offers a repelling force to either end of a magnet and appears naturally in bismuth. Offers a path for consciousness to enter the magnetic field control vectors [healing arts] and effect reality using intention.

Outflow / Inflow: A dynamic of the conscious energy to flow or establish flows into or out of conscious beings that exist separately from Source. These flows have an intelligence of a higher order, but also link us with the emotional and mental planes.

Source: "Source concept" the God force that powers atoms where we find the rest state of an atom is light speed, its rotational velocity. The control vector is the magnetic field and appears in two directions where spin is observed to be reversed. Proton and Electron, and merged in Neutron which contains dual spin.

Tempic: Coined by Wilbert Smith describes the light speed spin or motion that is fairly constant here in density 3 matter. When altered the light speed constant shifts slightly and sensation of time flow rate and gravity are effected. The tempic field is rooted in the Aether of the conscious planes and connects us all.

Disclaimer:

The devices are experimental and rely on the operators level of conscious interaction to manifest effects. These writings are based on our observations and perceptions and not to be considered "claims" or "prescriptions" for healing of for over unity devices. They are pointed at understanding our conscious level as mankind, and reaching for a greater unity through mental plane awareness. It is my personal goal to reach a state of "comprehension" while experimenting with these devices, as should be any who embark on the path. Only from this level can they be utilized to effectiveness.

Many thanks to all the participants on the c_s_s_p group for there input.
1 - 21 – 2007

Magnetic Vortex Generator - Construction Detail

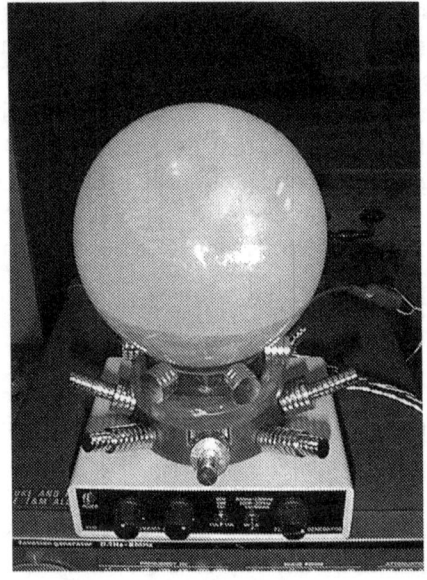

Aluminum Bismuth Scalar Coil

This coil is the heart of configuring the device for a Proton magnetic outflow or inflow.

There is much leeway in the coils, they can be custom sized to fit the device however try to keep the diameter under 1". 3/4" is optimal and bigger is not better. If a crystal ball will be used keep the coil shorter then the ferrite ring so the ball will sit on top.

Materials:

Aluminum tube - 3/4" diameter cut to 1 1/2"
Bismuth shot BB's [enough to fill the tube about twice there will be some waste]
24 gauge copper or tinned copper hookup wire about 20'
Electrical tape grey or red as black is rather ugly
1 Pkg modeling clay
1 small cast iron melting pot
Aluminum foil
Cookie sheet or other small pan
1 Propane torch - [or stove top can be used burner on high 520 deg F]
1 hot pad mitten for pouring
Safety glasses

Procedure:

The coil is made from an Aluminum tube 3/4" outside diameter 1 1/2" long. It is filled with Bismuth. Using a small pan cover the bottom with Aluminum foil. Set the tube upright and then seal it on the outside with a thick ring of modeling clay to contain the bismuth from escaping as it cools. The Bismuth is shot BB's that can be found at a reloading supply store. It is melted in a small cast iron melting pan that can be found at a kitchen supply store. It takes about 520 degrees F to melt the Bismuth, I used a propane torch. Bismuth expands as it cools so do not fill the tube completely. Take your time in the melting and slowly swirl the pan to get all the shot melted. There will be some scum at the top but as you pour the liquid bismuth the scum will be the last thing out so do not worry about it. Let the core cool well before touching it! The best coils will crystallize along the top of the coil as they cool and heating the metal longer and hotter will help this.

Next Wind The Coil:

Coil winding Detail

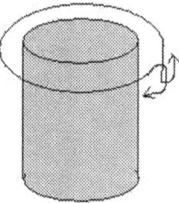

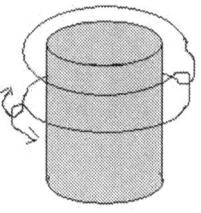

*Start in the middle of the 20 foot wire.
Loop the core at one end.
Twist the wires around each other once so
that wires trace back the way they came.*

*Repeat this on the opposite side
each time down the coil keeping
windings tight and as close
together as possible.*

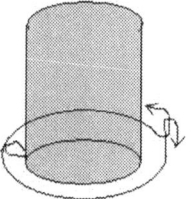

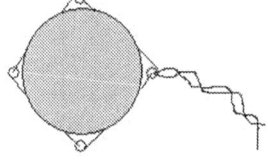

*After the first layer is completed wrap
the coil with one layer of electrical
tape and then wrap another layer
carefully up the coil.*

*This time though place the twists at 90
degrees to the underlying layer.*

*The finished coil should look like
this now from the top.
Carefully wind the two leads
about 1 turn per 1/4 inch and cut
at about 2'. A layer of electrical
tape over the outer layer will
keep the wires in place.*

Ferrite Ring

The ferrite ring can be scavenged from an older computer
monitor, the bigger the better. It is a slow job and care must be
taken not to break the ferrite as the chalking is chipped off near
the bottom. The ferrite ring sits inside the yoke on the back of the
picture tube. If you do not know anything about TV sets and
discharging the lethal voltages get someone who knows as the

voltages present even in a powered down set can be lethal. Be sure to keep the metal clips that hold the two halves of the ferrite ring together. If a ferrite ring is not available a 2 1/4" or 2 1/2" diameter iron pipe nipple can be used about 2" long.

Magnets

Many patterns can be used to set up the magnetic field however Neo magnets are recommended of sufficient quantities for a strong outflow to be achieved.

Here is my latest recommendation:

24 each type DC2 3/4" x 1/8" disc [placed around the bottom in six locations of 4 each stacked]

24 each type DA2 5/8" x 1/8" disc [Placed around the top in six locations of 4 each stacked]

Optional magnets can be added to these of a smaller 1/2" x 1/8" size to expand the fields width. D82 is a good choice.

If a large crystal ball is used this will expand the field envelope. A good magnet layout may cost around $60.

Use a compass to make sure magnets are placed correctly. North pole should be out on all the magnets and no reversals should appear anywhere along the ferrite ring. Depending the size of the ferrite ring it could take more or less magnet stacks. There will be a maximum limit of magnets before they start to push one another off.

Ferrite Coil

Function Generator Coil:

The Red coil around the ferrite ring shown above is optional and offers a scalar interaction with the iron atoms. It is used to create an Electron inflow or outflow vibration using a function generator to excite the field. It is made by laying red electrical tape turned sticky side out along the ferrite then wrapping a twin lead 24 gauge wire around it between the magnets as thick as desired then wrapping the tape over to hold the shape of the ferrite so the coil can be removed easily. To configure this coil for scalar operation the wires from each end of the coil are fed opposite directions for a magnetic canceling effect. As an option one end can be shorted and taped then the coil is fed from the other end.

Flat Pancake Coil:

The green Tesla type pancake coil has now been shown to be effective and useful as well for an Electron activation element in the device if one is researching power generation effects or just wants to feel the Electron heat energy. It easily throws out a 6 foot field around the unit with no bismuth coil inside just from the magnet, iron, coil interaction so lends to creating an Electron vortex device. This field is comfortable to work around and has no ill effects I have found as with the Lakvolsky and Mobius coils that produce chaotic energy. The crystal, if used, may be turned so that the lattice forms a V to the center of the unit rather then flat because this coil produces a conical field.

> About 24 feet of #12 solid copper wire
> Electrical Tape

SPDT switch [optional]
8 feet hookup wire [optional]

The coil is #12 solid copper wire with insulation, two wires stacked one on top the other. I placed 9 turns on this coil but the winding could extend fully to the ends of the longest magnets for the greatest effects. The coil is wound from the center of the wire, If you want an on off switch cut the two wires apart and solder them to smaller jumpers for extending to a switch, if not, just bend it over at the center lay it flat on the ferrite ring and start the wrapping process. Tape the next layer at each quarter turn and keep both stacked very neatly.

When you get to the outer end of the coil strip 1/4 inch and tightly tape the ends together so they are touching, or solder to smaller wires for on off switching. Both ends of the coil must be switched on and off using a DPST switch other wise either end will act to cancel magnetic fields.

Radiation pattern is a conical shape off the precession angles of the magnetic poles of the iron atoms and the Neo magnets and is found to be quite hot.

Function Generator

A program is available on the Kosol Core Tech group that has been especially designed to drive the coils using a computer sound output.

If this is used an 8 ohm to 100 ohm resister must be placed in series with the coil to protect the sound card of the PC.

Other wise a function generator can be purchased to drive the coil. Ideally one can experiment with frequencies, however NMR resonance effects happen between 1 and 20 Mhz for the metals present. Frequencies even as low a 8 Hz have been used, but I recommend the higher frequencies. As one works with the frequencies they normally tend to increase them over time.

Crystal Ball

The crystal ball pictured is a 125 mm Diameter Calcite sphere.

Quartz crystals have a natural electric vibration and any spheres can be used.

ZPE Coils

Preface

Intended for the betterment of mankind, these coils are the result of many working in the alternate energy research field of public domain information. This is a collective project shared by many sincere and giving individuals, with the intention of discovering a new source of power and healing for our planet.

Definitions:

Lathem coil - a coil based on a special length of wire [44 feet 6 inches] found to resonate a free torsion energy that has been discovered to be present in many places on the earth. They are typically wound into alternating scalar canceling and normal clockwise or counter clockwise forms, and wired in series to produce healing devices that are self powering to a mass of crystals present with them.

Terra coils - a coil based on the long side of a triangle distance crossing four Terra lines at opposite corners. The distance can be found experimentally by "pruning" but is theoretically about 45 feet 3 inches / fractions or multiples of this length to capture wavelengths or harmonics. As earth points may vary. Terra lines may be palmed or sensed in a local setting and various lengths can be connected to the RainMaker coils to sense the power levels of the captured fields.

Kosol coil configuration - One scalar bismuth coil is parallel wired to one normal wound coil with an iron content and both are setting in the same alignment. This system provides a strong linear torsional powering system. If placed into platonic form can create a density sphere system.

Limax Coil - Coil developed by Martin Pott, is wound with a floppy disc type flat ribbon cable. Ends are then spliced alternating to achieve a reverse current flow in each wire moving up the coil. This coil can be opened with a switch to shut it down. Coils can be now wrapped quickly to almost any depth.

Weave Coil - Designed by David Lowrance based on the description of Wilbert Smiths writings, a coil with each single loop reversed from the one next to it. Loops are placed on the coil form then twisted around one another at each of 180 degree points along the sides of the coil moving upwards. Normally a one layer coil is plenty for a bismuth core to become notably strong using a function generator. This is the coil detailed in the RainMaker vortex generator.

Tempic to electric conversion coil - Two coils of normal wind in non resonant lengths can be set up at 90 degrees inside a Faraday cage and tuned to receive torsion field waves induced using mobius or other scalar canceling coils. The coils must intercept the torsion waves at specific angles as torsion waves are polarized much like light or photon waves. Measuring equipment can then be attached via coaxial cables to monitor the presence of torsion induced waves. It was discovered during tests that many electronic devices actually radiate fairly strong 60 cycle torsion waves not detectable by normal EM equipment, and this is a common practice used to cancel unwanted EM.

Platonic form - Grouping of magnets or Torsion coils such that all torsion fields are equalized in all three dimensions forming a smooth confinement of space in which a higher density may be contained. Searl disc is patterned as a 2 dimensional platonic form. Kosol spheres as three dimensional platonic form. RainMaker base is a 2 dimensional form and the spherical crystal on top pulls this into a three dimensional one. Most all tube devices are single dimension.

Linear forms - Any device creating a strong torsion field in only one dimension, must be balanced with "mass of crystal" to avoid possible torsion sheers. This includes most all tube devices as determined by direct experiment.

Clockwise coil - A coil wound such that energy will move through it in a clockwise rotation as seen from the energy looking forwards to where it is heading.

Counter clockwise coil - A coil wound such that energy will move through it in a counterclockwise rotation as seen from the energy looking forwards.

General

It is discovered that there are two ways to wind coils and both produce different atomic interactions as to Tempic, Electric, and Magnetic field forces.

To determine wind direction of a coil, sight down one end of the coil and pretend you are moving into the wire closest to you and moving away. If you end up turning clockwise then you have a clockwise coil. Note that if you flip the coil over and do this again you still have a clockwise coil. Both coil winds are unique in this respect.

A clockwise wind will favor Proton spin, as energy moves through it, Proton spin is added to, and Electron particle spin is lowered. It is expected that this would be reversed in a counter clockwise coil. Noteworthy also is if Electricity moves through both coils it will produce a magnetic field in each that is reversed due to the Right Hand rule for magnetism. Proton spin results in torsion fields and Electron spin results in magnetic fields, so each coil wind alters both forces differently.

Energy In Copper Coils:

It is possible for all three forms of energy to move through copper.

Tempic - Torsion fields from Proton magnetism
Electric - Electron flow at electron shell
Magnetic - Electron shell in motion

With Terra coils we are working with the Tempic field and not the Electric field. To produce electric flows, some means must be used to reestablish the EM field from the photonic or torsion interactions. Tempic fields do not cancel but expand when set against one another, and this is the means of manipulation used to our advantage.

A Torsion Capturing Coil

This coil is basically a dead coil as to perception however within it is stored, invisible, a resonant torsion force much like a battery stores electric energy. If used in conjunction with other similar coils at 90 degrees to it or with other scalar coils the Torsion can be manipulated and becomes strongly perceptible. Altering the coils "wind" will alter its Torsion output. It is a linear form so is not normally used alone.

161

This coil is constructed from a 1/4 inch bolt 2 inches long, 44.5 feet of Grey insulated 24 gauge copper wire and a short length of electrical tape to keep the final wraps tightly held in place. Wind the coil along the bolt thread such that it follows the wind and a clockwise coil will result. It is then carefully wound to many layers back and forth to complete the length.

When connected directly to RainMaker this coil powers the unit to head popping power levels and the crystal sphere becomes too intense for prolonged use, showing the actual Torsional potential of the coil. The RainMaker unit can be used as a sensing device using this method, to determine the torsional state of coils and coil combinations if one has developed a working connection to it. This requires that the weave wind bismuth coils be used or some other non resonant length coils in the base.

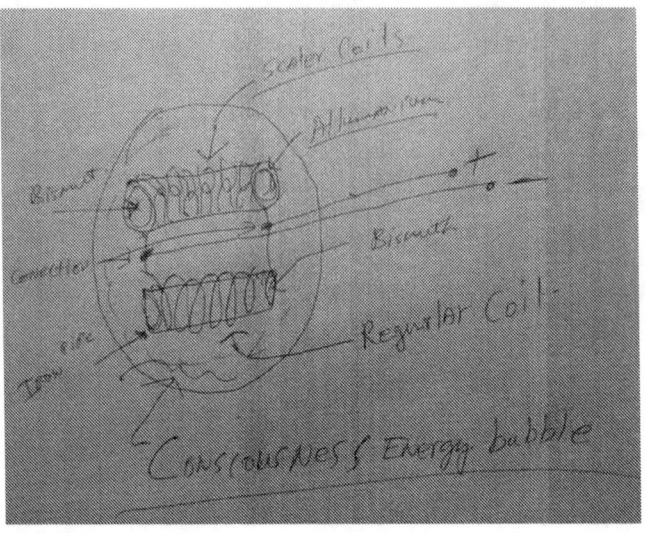

This combination as diagrammed by Kosol in parallel configuration, if placed into the inside of the RainMaker base will also produce very powerful Torsion fields, and in the case of my large 6 inch radius sphere are too powerful to leave operational for very long as they are in linear form.

A Simple Donut Coil

This coil will demonstrate to one the nature of the earth torsion fields present and allow for experiment on two dimensions of torsional field compression.

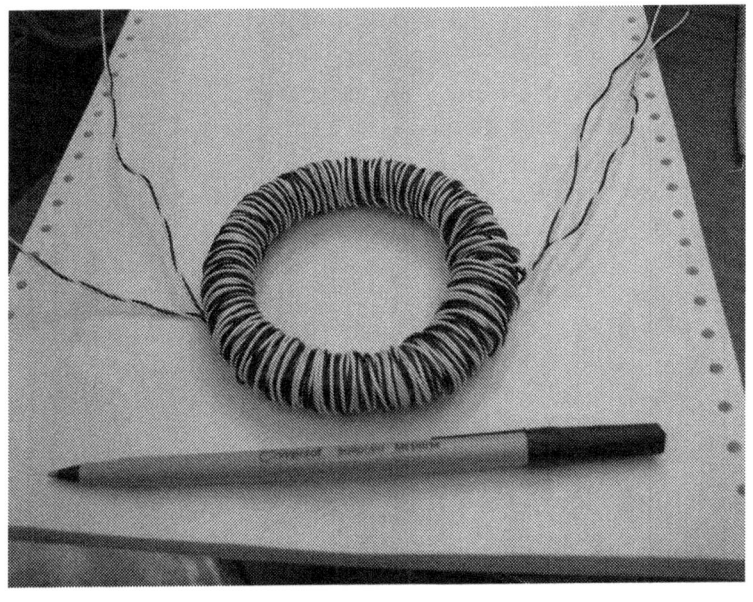

Construction is done using two torsion resonant lengths of twin lead 24 gauge wire. The first is wrapped around a 20 Oz plastic coke bottle. The bottle sides are now compressed to remove the coil and the second coil is wound around it to form a donut. This is done using clockwise wrapping techniques [check description above for clockwise coil]. Second coil can be wound from the center of the wire passed through the first coil so shorter ends will be pulled through on each wrap, also a clockwise wrap.

We now end up with 8 wire ends that can be configured to show the energy interactions. Inside is one black and one white horizontal wound coil and outside is one black and one white vertical donut shaped coil. The two dual coils interact at 90 degrees to one another and can create torsion fields in 2 dimensions.

To configure either coil as scalar canceling simply take opposite white wires and connect them to opposite black wires reversed, so the energy moving through the two coils passes opposite direction through each. Black to white in each case from opposite ends of the coil are shorted to form a scalar coil that can be shut down by opening the leads. Since the torsion resonant coils can be very strong this wiring technique saves one destroying the coils if things do not proceed well.

The most notable effect produced from this coil is not what one would expect. If the black and white twisted wire ends are connected together as a single coil in each case and not in scalar canceling mode, connecting one end of the horizontal coil to one end of the vertical coil produces a tempic field receiver. This is sensed as a feeling of "joy" by myself. If one now holds the other two ends in a hand touching skin, but not touching one another, the coil will begin to receive vibrations from any elements placed inside it. I have never used a Witness well or a stick pad, but this setup seems to offer a sensing effect for feeling the NMR rates of different forms of matter.

A Scalar Canceling Laythem Coil

A two dimensional coil form, if placed on a washer will focus most energy off its sides. If used in weak form will offer a standing torsion manipulation. It offers ease of winding and a quick access to torsion fields if control is not desired or necessary at low output.

The Toroidal coil shown above was altered in my designs to maximize power output by winding it on an iron tie wire coil form 28 feet 8 inches long and wound into the donut shape. [Lyle winds them on washers in the Star Link System].Then the Lathem coil is wound over this after folding the 44.5 foot of 18 gauge wire into four lengths of 11 feet. The 11 foot group of wires is pulled through the ring to its center position and then wound both directions. This speeds coil winding considerably for the toroidal shaped coils. 3 inches of excess wire then is left for connections [the red wires shown above].

The coils are quick to construct and powerful, but cannot be shut down easily, so expertise is required to balance the system using proper crystal to copper ratios.

The Density Sphere

Built by David Lowrance on 2/28/2007 and released to the public domain as no longer patentable.

A full function density generator offering torsion on all three dimensions as pioneered by Kosol Ouch. This can be adjusted in power output by turning its Earth alignment. The three scalar canceling coils are Lathem length coils but offer a control for shutdown not present in the Star Link systems, that was offered in the RainMaker ferrite base units early on. So this generator can be used as a stand alone controllable device. It is suggested that other coils be worked with first to determine whether these coils lengths are adequate for your locality. Ideally you will have your own custom lengths determined before constructing a full Density Sphere, otherwise the results can not be guaranteed to be consistent.

This device can fully replace the function generator used on RainMaker units and should not produce any notable headaches of itself at high power levels. The strong fields may however be altered as they are used to power other devices, which can result in a linear compression, so care must be taken in design of the powered devices as well.

Construction

The forms for the wire are constructed by cutting cardboard rings and then taping these rings to form a full 3 dimensional coil winding container. Sphere chosen is 3" copper laced quartz having a rather strong field to begin with. Sheet rock knife, scissors, and various tapes are used. It is desired that the crystal be free to turn inside the unit later so care is taken in construction.

Rings must set on the sphere slightly to one side of center, but is not too critical as another ring is placed directly on the sphere to separate the rings.

Showing the six rings necessary for all the materials to hold the coils on.

Fold the rings in half and trim so they loosely fit over the sphere on a quarter section.

Cut lengths of cardboard also now to seat under the rings. It must be custom sized to fit the coils wire amounts for width and sphere size. A coil can be wound ahead of time to estimate thickness. Coils can be any torsion resonant length or multiple resonant lengths for your local earth resonance. The strips are fitted then taped to cover the sphere on all three planes of spin such that the crystal may still be moved inside them for alignment if this is desired.

The rings are now cut into quarter sections.

Quarter sections are lined up on the crystal and folded to form into triangle shapes then taped into place using electrical tape.

No tape contacts the crystal surface but is contained on the coil form such that the crystal may be realigned inside the coils after construction.

Showing completed coil form around crystal sphere allowing 3 coil winding containment channels. The unit is set in the top of a tape roll in this case to keep it stable while tapping the last pieces into place.

More tape is added to seal all joints and provide a clear surface for wires.

Coils will now be wound in layers clockwise wind, one layer over the other in each quadrant until the complete wire lengths of twin lead 24 gauge 44.5 foot lengths are all used.

All three lengths must be started simultaneously, then the lowest quadrant is worked one at a time on each layer. The coils are wound by taping one end of all three wires in place at starting position and working one spin plane at a time for one layer. This will produce three interlaced coils of twin lead.

Connections

The coils are all now configured as scalar canceling by joining opposite ends of opposite colors on each one. Next the scalar coils are all three wired in series to provide a total copper interaction that flows through all of them. Two of these can be opened to be used as outputs to the RainMaker unit for direct sensing of the energy inside the Density Sphere.

Adjustment And Control

Align the sphere of the Density generator so one coil is horizontal and the other two are about 45 degrees from the North Geomagnetic pole of the earth.

As you turn the coil now along its horizontal surface the intensity of the field will be altered that is emitting from the RainMaker and can be felt at any angle coming from the RM sphere. Direct input can be sensed also by touching the Density Sphere with finger tips on its crystal surface.

Coils can be deactivated by opening all the opposing coil leads and disconnecting from the RM unit.

The energy produced from the Density Sphere is "fun" and would seem to contain a high frequency healing crystal component vibration. At present I would class it as angelic female energy.

Just remember the scalar coils must by 90 degree from siniod coils, or vice versa. That means siniod coils must be 90 degrees from scalar coils in order to create zero point energy torsional force.

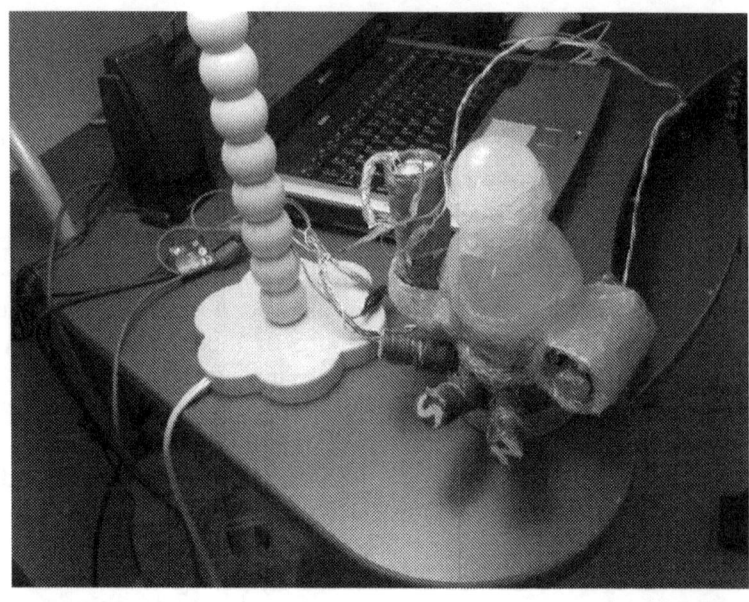

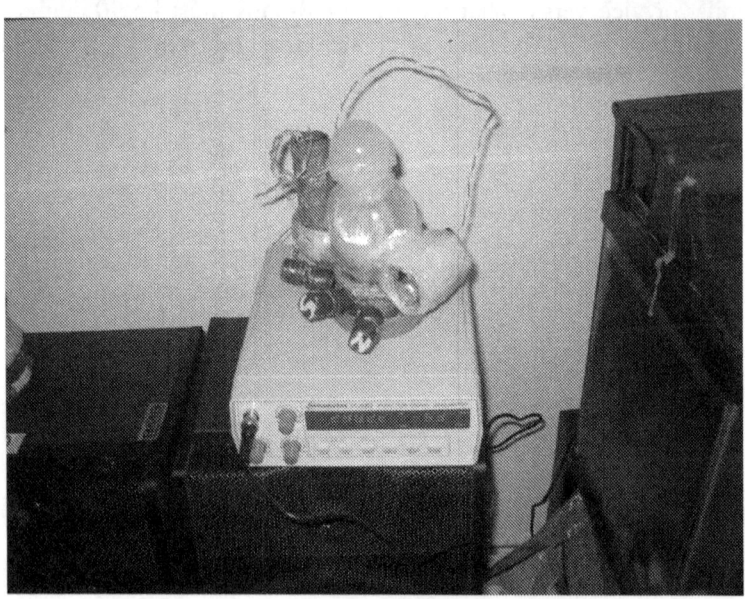

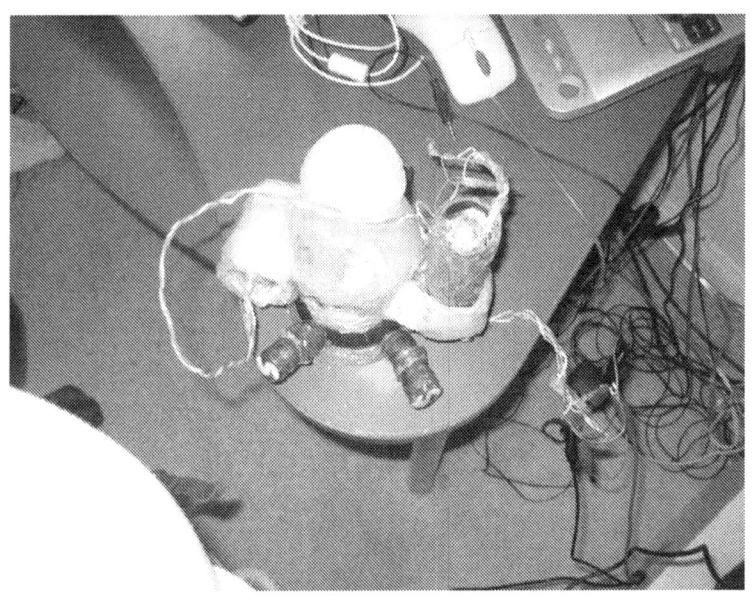

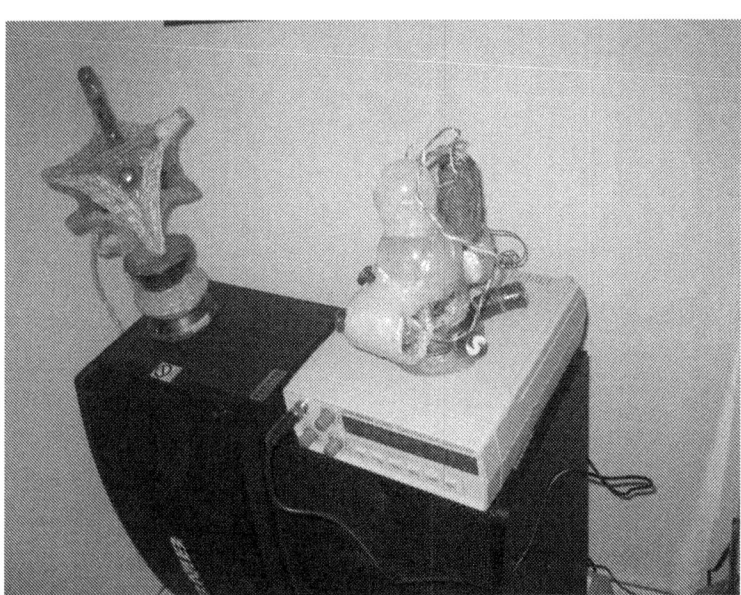

Helix Coils And Scalar Coils

Scalar coils or bismuth scalar coils are made from DNA/RNA proteins pattern. Human beings and plants and crystals are scalar entities just like scalar coils.

Scalar coils are created from the pattern of the DNA/RNA protein sequence from our body. The DNA/RNA of a human being is made of charged protein, this charge are fields which cancel each other and as result produce torsional fields. Torsional field are attracted to each other and will build up so any coils that are built to the pattern and physical structure of DNA/RNA sequence protein pattern form. This allows such coils to interact with and promote that biological entity through its charged fields alone so interaction and symbiotic occur. This connection of fields is very very physical to the sensation of the organism between any object or beings that is in the presence of that scalar coils that is emitting the fields.

This sensation from fields of the coils to the fields of the DNA/RNA human or plant or animal organism become symbiotic and experience is so real that the addiction occurs because of the building up of fields through interaction symbiotic. The scalar coils or canceling coils is another name for DNA/RNA coils is created from the observations of DNA/RNA protein sequence. DNA/RNA is called scalar or canceling pattern and this DNA/RNA do produce scalar fields also, just like the scalar coils. The scalar coils were created from observing the pattern the DNA/RNA arrange.

The source of this intelligence is from a super alien intelligence who connected the dots and guided the way to create these kinds of consciousness scalar coils which mirror the human DNA/RNA protein pattern.

Human beings and plant and animal as well crystal is call scalar life form, so therefore is able to interact with scalar coils or cancelling winding coils.

Regards,
Kosol Ouch

And also here is from my friend Thomas C. Kramer on the helix coils or scalar coils.

New Physics And Old Chemistry
September2007

When I went to college Physics and Chemistry were two different subjects. In Physics I studied waves and forces. In Chemistry I studied atoms and their interactions. But then I met a Bomoh, an Asian shaman, who defied all the laws of physics and chemistry by making solid objects appear out of thin air. For someone with a strong scientific background I was amazed and first looked for trickery and hypnotic suggestions or other worldly explanations, but all failed. So I have spent many years trying to understand how solid objects can be transformed into and out of our dimension.

The simple answer is that there is just a change in frequency.

It is something like changing radio or TV channels. You just change the frequency.

But it isn't quite that simple. But we do know that everything vibrates. Even at absolute zero there are still movements. But is this a vibration like we think of when measuring a sine wave or one like we observe when plucking a guitar string? Or something else?

In my studies of electronics I have found that electrons do not move through a wire very fast. What we call electricity is more of a field effect that moves along the outside of the wire creating an electro-magnetic field perpendicular to the wire all around while the electrons (?) are made to flow in one direction (DC) or to constantly reverse directions (AC).

But what is an "electron"? Physicists and chemists still haven't figured that one out. Some say it is a wave form (a photon), and others say it is a solid particle, while still others say it is both as it flip-flops between either state.

To me an electron has to be one or the other because flip-flopping is for idiots. So which is it and why? If everything vibrates then an electron must be some form of wave form and not a solid thing. If we go back to the transmission wire above, the electro-magnetic field is all around the wire and the electrons are moving on the outside of the wire all around as they come off the end.

This appears to me to be a spiraling movement and not a sine wave form. That is, the electron (?) waves spiral down the wire at a high frequency.

When we apply a force to one end of a wire it acts much like the old 5-balls on strings experiment. When the first one is knocked the ball on the opposite end is released. But what is released?

Regressing a minute, I would like you to think of an electron wave as being like a spring, a long coiled spring. Now we know that we can stretch and compress a spring which will make it longer or shorter without changing the diameter (frequency change at same amplitude). But we can also twist the spring that will have the effect of making it larger or smaller in diameter (frequency same but amplitude changed). Or we can bend the spring and thus change its direction. And of course, we can do all these things at one time.

Now if an electron is a wave form, we know that it can be compressed or expanded, twisted by electromagnetic forces and bent so as to change directions. Simply we know how to alter frequencies, amplitude and how to bend light.

But why do electrons appear solid and where do they come from?

Firstly it is necessary to understand the concept that we are existing in a huge energy soup mix of different spiraling wave frequencies (springs of a myriad of sizes).

We also know from chemistry and physics that solids vibrate at fairly low frequencies, liquids generally higher, gases still higher, then we have the influence of sound waves, microwaves, infrared, visual light, ultra-violet, x-rays, gamma rays.......and who knows what else......electron waves? All spiraling along at different frequencies, amplitudes and directions.

We also know from music that various frequencies are in harmonious chords or discords. Electronics have the same effects. Keeley and others played with these tunes and have done amazing things like levitations, weight increases, free energy acquisitions and many more.

We also know about tornadoes, hurricanes and whirlpools and the powerful compression and vacuum effects that these natural occurrences cause.

Now if we look at a typical dynamo or electrical generator what we see is a magnetic rotor passing by a stationary magnet, that is, one field cutting past another. If we do the same thing, passing one field past another, using air or water we create whirlpools (twisted springs that get smaller/compressed or larger/expanded by altering amplitudes).

Electrons (waves) magically appear as a result of these electro-magnetic fields passing by each other. To me this appears to be simply the creation of magnetic whirlpools appearing in the spring soup that we live in, that is, a compression or expansion effect on a wave (or standing wave) as a result of a friction-like interaction that occurs when two opposite fields pass by each other.

Now what would happen if I wound 2 identical springs together? You can easily twist one into the other and what you get is a "Double Helix"! Does that ring any bells? Perhaps one flows in one direction and the other in the opposite direction. Do we call this 'positive' and 'negative' flows? Perhaps DC or AC? N-S poles?

What if we add compression or expansion to these double waves? Frequency variations? Bending? What happens if we do this in harmony to other strings of springs or in discord to these?

I hope you can see where this is leading your understanding.

Now add 'intent' and mix a zillion waves together and you get "me"! Or YOU! Or everything in our current reality. But change the channel (reduce the amplitude and increase the frequency) and perhaps you enter another dimension.

Now how does this apply to free-energy, zero point energy, orgone, Joe Cells, electrolysis and physics and chemistry?

Back to the beginning... everything is based on waves. There really is nothing solid, only things that appear to have density based on their lower level of vibration and higher amplitudes.

If this is the case and we live in an energy soup then in order to tap into this energy and bring it into our level of reality or density we will require machines that are designed to alter these very high frequencies and small amplitudes. Simply we change channels by tuning into the harmonious frequencies of the soup and then amplifying them.

In electronics this is done with coils. But if we take a twisted pair of wires to make a coil, all kinds of strange things happen. Then if we make pancake or cone coils using twisted pairs still more strange things happen. (Tesla had fun with these). And I suppose you could use a quad helix (2-twisted pairs twisted together) or other more complex coil formations to get even stranger reactions. Then there are coils inside coils (Tesla Coils).

Now add a frequency pulse (spark gap) and see what happens! The lightning come on yet?

Barber pole or helix spiraled electrolyzers are coils that have the same type of interactive effect on water. Splitting water using electrolysis is clearly a surface effect and even better with a rough surface.

Now if electrons are "created" by passing fields, like passing electricity one way down a wire and the opposite way down an adjacent wire, you would expect electrolysis to work best in such an alignment.

I would think it would work even better with counter-rotating helix electrodes. In fact, weight loss has been measured using similar coil or counter-rotating field effects.

Tesla also discovered that there was a 3rd type of electricity (cold electricity). Simply in a 2-dimensional plain electricity spirals down the outside of a wire and creates a magnetic field perpendicular to that flow, but in a 3 dimensional world there is another field that is perpendicular to both the electrical and magnetic flows. This is the X-Y and Z lines of a 3D graph. The Z-line represents this 'other' energy field that we can tap and use.

The problem then becomes one of creating machines that can tap this Z-field and then use the energy so derived. This is what several inventors have done including Tesla, Morey, EV Gray, Reich, Joe and others.

The basic technique used is to collide a similar electrical flow (positive or negative) so that neither can reverse itself and then surround the occurring spark gap with concentric collectors noting that the Y-magnetic and Z-field will be collected but only the Z-field will be used as the X-electric and Y-magnetic are transposed onto the new wire carrying the Z-field. Sounds confusing but if you just look at a 3D graph and picture the X-, Y- and Z-fields you can easily see how you can flip one to the other. And that is simply what is happening in many of these devices.

Similarly, if such a system is tuned to be in harmony with the ultra-high frequencies of the soup we live in, then there is often a measurable increase in energy output of these devices. And alternatively, if these devices are in discord or 'jamming' the frequency soup, then there may be energy absorptions measured.

To tap into these fields is thus just a matter of choosing which axis field you want to use and then designing a 3D instrument to take advantage to that axis field.

A helix coil is a simple example of this. The twisted pair of wires is wound around a core. Power is applied so that electricity flows down one wire and then loops back on itself forcing the X-Y-magnetic fields to collide with each other thus affecting the Z-field.

If the coil frequency is harmonious then a positive power increase is noted.

The trick then is in the specific design of these machines so that you can take advantage of whichever field effect that you desire. This is what Tesla, Keeley, Morey, Gray, Reich, Joe and many others have done. A dynamo-generator crosses the Y-magnetic fields in similar polarities to create electricity, but the Z-field is perpendicular to the Y-field and is necessary for electric generation.

Similarly, magnetic fields are created from the X-electric field and the Z-field interaction.

So if we go back to the 5-balls on strings, we can see that if a force is applied to a linear string, it results in a ball popping off the opposite end and then coming back to pop a ball off the other end, and so on and on, so that there is really no movement of most of the balls in the suspended string, but only measured observations at the ends or at sensor points along any part of the string of balls.

This is where the "standing wave" theory comes in and our fine soup thickens in 3-dimensions.

Now jump into the SOUP!

Argh! There are electrons, positrons, micro-atomic particles, protons, anti-protons, neutrons, and these form all kinds of atoms which come together to form molecules which, which, which eventually make up you and me and this whole 3D world that we live in and the universe that surrounds us. How this all fits together is so complex that the mind cannot possibly comprehend everything in its entirety. I guess this is why we have a Creator after all.

But just thinking about creating yourself out of a vibrating soup is in itself a supreme act of willpower and that is what we really do. Amazing, isn't it?!

So, back to the beginning of how a true mystic can make solid objects appear out of thin air, it boils down to frequency and amplitude manipulation of a 3D plain. This, of course, leads one into the realms of the Invisible Man, ghosts, Jinns and other-worldly beings and can be further extended to "cloaking" and UFO theories and so on.

The late Dr. Bob Beck and Dr. Andrija Puharich both carried out extensive experiments on clairvoyants, trance mediums, shamen and mystics. Puharich discovered and promoted Uri Geller and the Brazilian healer Arroyo. And one of the things that they found was that when these talented individuals were in their "working state", they emitted very high frequencies. Mystics capable of physical transformations generated frequencies in the 1-100 kilohertz range (that is in radio frequency ranges!).

There are also the stories of some of the survivors of the Philadelphia Experiment who were able to walk through walls. And that has to be an induced frequency change caused by the high frequencies that were generated around the ship to make it "disappear"!

In the final analysis it appears that it may be possible to alter ourselves and the world around us by simply changing frequencies. For those of you who have read my "Invisible Machine" article you will perhaps understand how this can be done using simple electro-magnets and pulsed circuits. It has been done. And you can bet that the technology is much further along than any government is willing to admit.

But this understanding is and will be a very important part of our futures. We are bombarding ourselves everyday with higher and higher frequencies, EM fields and more, so we will change ourselves and our world. Perhaps this is what is part of Ascension. Perhaps it is just us getting smarter.

Anyhow, I hope that I have given you all some things to conjure and mull over.

TCK

Thanks Tom for contributing to the book.
Regards,
Kosol Ouch

Stargate Ascension Method

The stargate ascension method is now modified for perfection and can be used in conjunction to the vortex generators known as the RainMaker device.

The modifications of the methods go like this, first you just meditate on the visualizations of the souls seat sun which located between the fifth chakra and the fourth chakra of the 7 chakras system but in the 14 chakras system the souls seats is the sixth chakras. then after you do that for a good 10 to 15 minute then you just move the meditation focus to the center of the heads sun which is where the hyperthymus gland that contains the pineal gland and pituitary glands. If you imagine a line of light from one top of the ear to the other top of the ears and a line of light from the center of your fore heads to the back of your head then where ever the intersection is, that is where you visualized a sun that is the size of a golf ball, about three inches in diameter. Now you focus on this visualized golf sized sun as long as you like then you switch to the soul seat sun which is big as your fist. Then you continued on focused your meditation on the visualized sun of the soul seat, and switched between the center of the head sun and soul seat sun. Now you know you are in the half lotus position for the facilitator and your traveler are laying down his or her hand is touching your knee. so you breath just using the cooling breath in other words you breathe normally and softly, when you breath in through your nose say in, and when you breath out through your nose you say out at the same time your focus of attention is focused on the visualized soul seat sun and then moved on to the visualized center of the head sun. You repeat the process of breathings through your nose in and out with mental chants of the words in for inbreath and out for outbreath at the same time you mind is also is aware and focused on the center sun and soul seat sun by switching back and forth periodically. for example focused on the soul seat sun for about 15 minute then move that focus awareness to the center of the head sun for about 20 minute then move back to the soul seat sun again for 15 minute then center of the head sun for 30 minute etc. Back and forth as long as you like to extend your facilitation of the stargate travel to your traveler, but also all of this methods is to be used under a pyramids or the methods is to be used in conjunction with a RainMaker device known as vortex generators also this device give both the facilitator and travel super extra consciousness divine energy call chi, prane and also vital life force. This energy helps the facilitator and traveler to heal and also to travel interdimensionally to other plane of reality to visit loved ones and super cosmic alien civilizations in this plane and other higher planes of existence, etc. No limit whatsoever.

Please get a book that is written by me (Kosol Ouch) called *Star Gate Ascension The Cure to Boredom and World Disease* as well please get the book by me and Jerry Evans II called *Cultivating Inner Force And Reading People Like A Book*. Other books are *The Rain Maker Device;* the *Technical Guide for RainMaker Device, Ghost Consciousness Catching Device, Zero Point Energy, Ascension Machine and Over Unity Coverage*, and *Kosol and Koeun Noun Ouch Spherical Generator*. All by Kosol Ouch and his co-authors. You can get my books from Amazon.com, Borders.com, BarnesandNoble.com, etc.

Kosol Ouch

Kosol: Love Conquers All

The entire RainMaker device and stargate meditation technique as well as the light body technique are all designed for the Earth people to put into practice the science of consciousness and its technological tools to assist humanity and all sentient life to develop the light body as well to connect everyone to the akashic records fields with bind all things together from universe to galaxy to planet to every aspect of life form and sub atomic particle wave, etc. Love is the greatest wave of all because love has no limit whatsoever and is the truth of truth and love conquers all. Without love nothing exists and love is the true reality for both physically created reality and spiritual reality. This device and stargate meditation technique will help all beings to realize that through living the experience. No limit whatsoever. All of this consciousness technology and meditation technique uses love model and practicality of technology on a no limit level. So all beings will serve the divine love on a universal and planetary to galactic to solar system level as both galactic, universal and planetary guardians through all universe and dimensional reality.

Using Rainmaker Devices

It has 4 crystals, 4 geodes and a octohedron made out of florite in the center, it's covered with copper screening. It sits inside the iron rings with magnets on it.

Now the most important thing is that people who have built RainMaker devices must use them. You use them by putting your hand over the RainMaker device crystal or geo, rock, etc., and just feel the auric energy coming from the RainMaker crystal, geo, and rock, etc. Then you move your hand over it in any direction you want. You can use both hands as well. You can feel its auric energy fields from 3 inches to 3 feet or more, depending on your level of sensitivity. By playing and practicing with your hands over the RainMaker device rock, geo, crystal, etc., you begin to develop sensitivity to the auric energy fields of the RainMaker device. This in turn will allow you to begin to see aura as well as develop telekinetic and telepathic as well.

As you practice more and more you begin to develop light body as your spiritual body becomes integrated with you physical body. This is what we call the merging of spirit with physical therefore have light body, a body that doesn't die or age and that doesn't get sick. The Christians call it a new creational body or Christ body. Others call it light body or mekaba body as well the Buddha body. Now the best part of all of this is that you can practice 30 minute or more and then you can meditate around the device once you become sensitive to it. Then you practice both meditation around the RainMaker device and also feeling its aura. Practice will speed up your DNA/RNA from a 2 helix to a 12 helix DNA/RNA. The device will help activate your DNA/RNA and transmute it to a 12 helix DNA/RNA which means you have light body by just feeling its aura - practice and meditate around it. No limits.

So therefore have fun and enjoy your RainMaker device, a gift for humanity from the spiritual worlds and the galactic federations of light. As well, be prepared for first contact, before December 21, 2012. On December 21, 2012 Earth and its solar system will enter the galactic stargate and will move closer to the Sirius B star system. 8-28-2008 is the beginning of first contact. It will last for 2 hours. A massive galactic first contact fleet will arrive on Earth. People will have reunion with their space and spiritual family, as well the ascended masters of all fate Christ and Buddha, etc., will return on the day of first contact between humans of Earth and the galactic federations of lights. So this consciousness technology is a gift from the galactic federations of light and father/mother gods to help prepare Earth humans for first contact with the galactic federations and the divine.

The co-author and I will see everyone in first contact which will begin the process on 8-28-2008. So see you there.

Regards,
Kosol Ouch

Miscellaneous Pictures

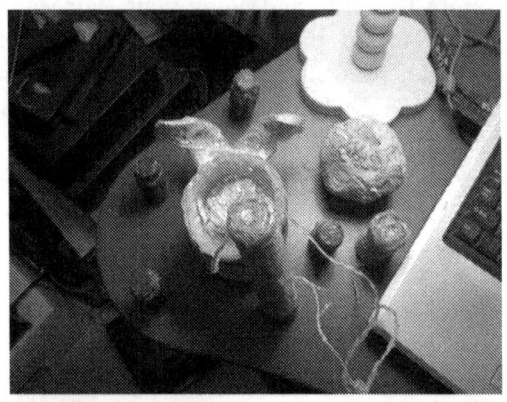

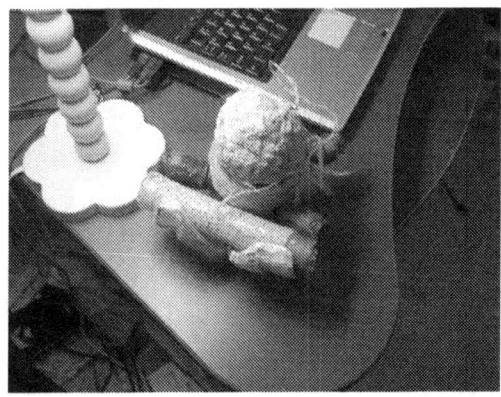

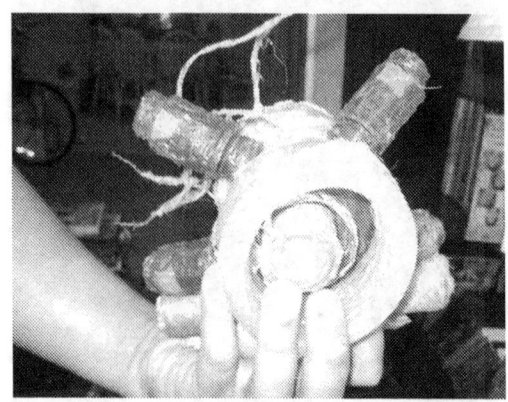

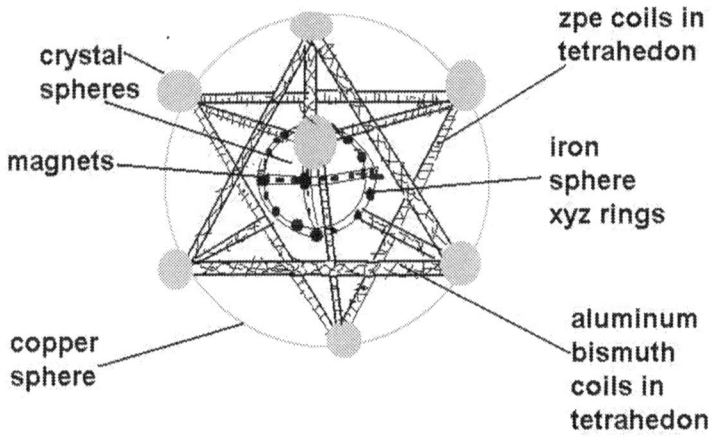

crystal spheres

magnets

copper sphere

zpe coils in tetrahedon

iron sphere xyz rings

aluminum bismuth coils in tetrahedon

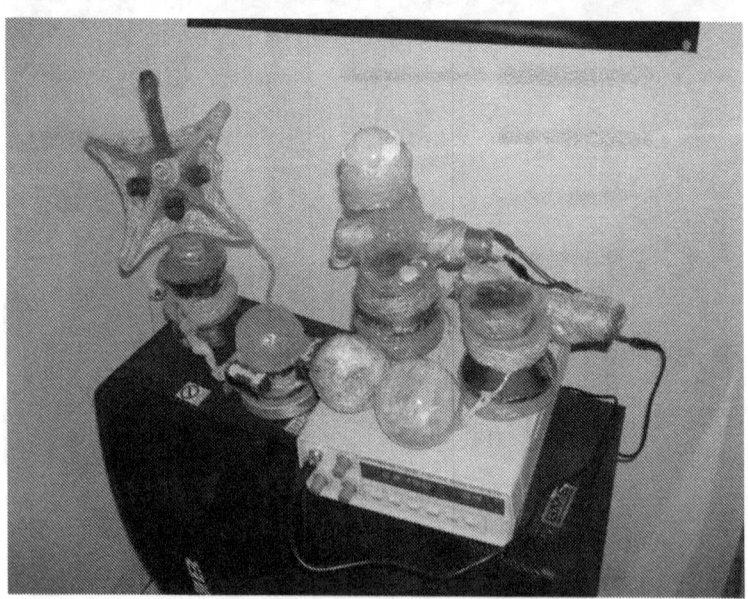